THE LOGOS OF THE BIOS 2

Bio-Communication

Published by Umweb Publications

www.umweb.org
info@umweb.org

Umweb logo by Dario Martinelli
Cover artwork by Wilhelm Hasenauer
Coverphoto by panthermedia.net/Heiko Kramer

Language editing by Natalya Sukhova

Printed by UAB „Biznio mašinų kompanija"
J. Jasinskio g. 16A, Vilnius - Lithuania
info@bmk.lt - http://www.bmk.lt

ISBN-13 978-952-5576-04-7
EAN 9789525576047

THE *LOGOS* OF THE *BIOS* 2

Bio-Communication

GÜNTHER WITZANY

UMWEB

Acknowledgements

I would like to thank Frantisek Baluska, Peter Barlow, Nikolaus Bresgen, Mario Gimona, Erich Hamberger, Randy Jirtle, Frank Ryan, Jim Shapiro and Luis P. Villarreal, for inspiring dialogs, Dario Martinelli and Natasha Sukhova for warm-hearted support and review, Michael Stachowitsch and Pierre Madl for translations, Wilhelm Hasenauer for great design and Hiltrud Oman for mental support.

Contents

Introduction

In the second half of the 20th century the physiology of all levels of cells, tissues, organs and organisms of all organismic kingdoms was the mainstream direction in biological research and experiments. In the 1970s increasing metaphoric use of "communication"occurred. In the last decade of this period the notion of communication (without being used any longer as a metaphor) within and between organisms passed the pure physiological understanding. Cell-cell communication dominates contemporary cell biology giving a deep insight about a great variety of signalling pathways for both organisation and coordination for production, release, uptake and information processing within and between cells.

In parallel the metaphoric usage of "language" has increased since the midth of the 20th century with the growing knowledge about the genetic code. Most of the processes which evolve, constitute, conserve, rearrange the genetic storage medium of DNA are terms which are originally used in linguistics such as coding, copying, translation, transcription, etc. Meanwhile also the linguistic approach lost its metaphoric character and the similarity between linguistic languages/codes and the genetic storage medium is not only accepted but tends to be adapted in bioinformatics, biolinguistics, protein linguistics and biosemiotics. The advantage of methdodological adaptation of communication and linguistic terminology is to have appropriate tools for differentiation at specific levels which are difficult to describe non-reductively by pure physiology.

This means that language-like structures and communication processes occur on the bottom of living nature and are not at all evolutionary inventions of humans, nor are they anthropomorphous adaptations of terms of description of human capabilities on non-human living nature. It simply became and still is obvious that every coordination and organization within and between cells, tissues, organs and organisms need signs, i.e., chemical molecules which serve as signals or symbols in messages or serve as vital indicators for environmental conditions. These signs should be interpreted in a correct way by biological agents, i.e., there must be subjects/representatives of sign production and sign interpretation. Thus, the interpretation may also fail resulting in the lack of coherent or appropriate response behaviour.

The *Logos* of the *Bios* Vol 1 and Vol 2 focus on this change from a viewpoint of a pure mechanistic biology to a viewpoint of a linguistic, semiotic, communicative biology. Volume 1 gathered articles which were published between 1995 and 2005 in several journals. The aim of these articles was that a linguistic, communicative, hermeneutic or semiotic view on biological phenomena needs the integration of an non-reductive description method of sign(alling) processes (semioses; Gr. Semeion = sign) which is able to avoid classical one-way discussion in the theory of science between 1920 and 1980.* Therefore Vol 1 suggested contributions to a foundation of a three-leveled biosemiotics, i.e., that any investigation or analyses of sign-mediated interaction has to recognize the complementarity of syntactic, pragmatic and semantic levels of rules. This is the consequence of the fact that all former trials to detect the illocutionary sense of utterances (respectively deep grammar contents hidden in surface grammars) by a pure linguistic (syntactic/semantic) analyses failed, because it is not possible to iden-

* They are abstractive fallacies like methodological solipsism (sender-receiver narrative; subject-object dichotomy) and its consequence such as objectivism.

tify the pragmatic rules which constitute interactions by investigating syntactic and/or semantic rules.

Without sign-mediated interactions no coordination or organization

In Volume 2 the *practical application* of the early developed pragmatic philosophy of biology follows. Articles from 1 to 5 are reviews which cover all organismic kingdoms in special examples, except that of Archaea. The aim was to demonstrate Bio-communication in all domains of life as rule-governed sign-mediated interactions.

Bio-communication is the precondition that coordination within and between organisms can take place. Bio-communication occurs on three levels: (A) intraorganismic, i.e., intra- and intercellular; (B) intraorganismic, between the same or related species; and (C) metaorganismic, between organisms which are not related and in most cases are of different organismic kingdoms as happens in most symbioses.

The pragmatic philosophy of biology, which I developed between 1987 and 1990,* poses, that under the assumption that life is structured linguistically and organized communicatively, there must be a complementarity of three levels of rules wherever signs are used within (inner nature) and between (outer nature) living organisms. Because one sign is not a sign, and one sign user cannot use signs without a community of sign users which he/she/it is a part of (may it be by socialization, i.e., memory and learning, or by innate competence transfer), there is a close relationship to the same or related species which share (A) syntactic, (B) pragmatic and (C) semantic rules of sign use, i.e., how signs can be combined or sequenced (syntactic rules), how interactions like coordination and organization can be established (pragmatic rules) for different real life purposes like common behavior,

* „Natur der Sprache – Sprache der Natur. Sprachpragmatische Philosophie der Biologie"; 1993, Würzburg: Koenigshausen & Neumann; translated und updated 2000: "Life: The Communicative Structure", Norderstedt: Libri Books on Demand

or attack, defense, mating, virulence, nutrition uptake, etc. Dependent on the pragmatic context are the semantic rules which determine the meaning of signs, i.e., how one syntactic sequence can have completely different meanings in different situational contexts.

The pragmatic philosophy of biology also proposed, that different levels of semiotic investigations must be possibly similar to zoosemiotics to describe the communicative nature in "plant semiotics, fungal semiotics, protoctist semiotics and a prokaryont semiotics" (Witzany 1993, 2000). In this Vol 2 this proposal is realized in the more general biosemiotics of all organismic kingdoms (Chapters 1-4).

Three kinds of signs: Index, Icon, Symbol

According to the founder of Semiotics Charles Sanders Peirce we are able to distinguish three different kinds of signs, Indices, Icons, and Symbols. Indices are in most cases abiotic stimuli from the environment which are interpreted as signals, e.g. a plant root identifies nutrients as being relevant as the plant shoot does with the angle of sunlight. Icons are biotic one-to-one signals (analog) which need no further explanation, e.g. plant cells identify auxin in a hormonal coordination process. And there are symbols, i.e., signs or sequences of signs like characters of an alphabet used to generate words, sentences, codes which do not indicate by themselves what they mean (what their function could be) but are signs through natural or cultural conventions. Such sequences may also be sequences of behaviors as the dances of the honey bees of the northern hemisphere (see footnote at page 8) or even the eukaryotic genome grammar with its high complexity of alternately coding and non-coding sections.

Life: The Communicative Structure - pragmatic philosophy of biology

FORMS OF COMMUNICATION

TYPES OF COMMUNICATION

living nature: linguistically and communicatively structured and organized	**PROKARYOTES** bacteria	**PROTOCTISTS** eucaryotic microorganisms	**ANIMALS**	**FUNGI**	**PLANTS**	self-understanding of **HUMAN BEINGS**
intraorganismic communication (pragm., synt., sem.)	protein synthesis; feeding	protein synthesis; metabolism; true nucleus, cell respiration; chromosmes	protein synthesis;metabolism; digestive system; organs; nervous system; brain; consumption	protein synthesis; Metabolism; germinating spores; mycelia; absorption	protein synthesis;metabolism; photosynthesis	similar to primates; expanded brain structure
interorganismic communication (pragm., synt., sem.)	association; simple colony formation; coordinated behavior	association; more complex colony formation; coordinated behavior	complex forms of coordinated behavior; greatest diversity of sign use; forms of social interaction; precursory ritual behavior	association; formation of large colonies; sexual reproduction; coordinated behavior	association; colony formation; high diversity of chemical linguistic signs;	generating of cultural norms; social life-forms; communication via speech and writing; religion, politics, science, fine arts, crafts
metaorganismic communication (pragm., synt., sem.)	parasitic; symbiotic	parasitic; symbiotic	parasitic; symbiotic; social; phenotypic attack and defense strategies; mimicry	parasitic; symbiotic; attack and defense strategies;	parasitic; symbiotic; attack and defense strategies;	parasitic; symbiotic; technical exploitation of all kingdoms of organisms; responsibility for the survival of life

Figure 1: Types and Forms of Bio-Communication (Witzany, G., 2000)

Context determines meaning

For chemical molecules which serve as signs are not indefinite, it is normal that sign using organisms share a finite number of signs and a finite number of semiotic rules which are necessary to use. So it is common to that living organisms that they use the same chemical substances (in either liquid, gaseous or mineralized forms which serve as signs) to transport messages in different pragmatic (situational) contexts; Auxin in plant communication may transport various messages whether it is used in the context of neuronal-like cell-cell communication or as hormone to transport messages between root and shoot or as morpogenic sign. Some dances of the bees serve as symbols to transport the message of an appropriate hive *or* in another pragmatic (situational) context to transport the message of an appriopriate place to find nectar. Not only in that case, the specifical geographical situation of the populations for growing up and socialization may lead to special dialects, i.e., the signs which are used to communicate are identical but the meaning of the transported messages differs dependent on regional customs. But also this phenomenon of dialects we find in bacterial communication. Dialects in bio-communication could be an important feature in the investigation of in vivo eco-fields as suggested by Almo Farina.[1*]

---

[*] Farina A, Belgrano A. (2006). The Eco-field Hypothesis: Toward a Cognitive Landscape. Landscape Ecology 21: 5-17.
Farina A, Scozzafava S, Morri D, Schipani I. (2007). The eco-field: An interdisciplinary paradigm for ecological complexity. In: Witzany G (ed). Biosemiotics in Transdisciplinary Contexts. Proceedings of the Gathering in Biosemiotics 6. Helsinki, Umweb.

Living nature and non-living nature

The difference in pure stochastic/mechanistic biology is that semiotic rules of sign use cannot be explained sufficiently by natural laws of physics and chemistry alone, because in contrast to natural laws such as gravity, living organisms are in continuous relations to semiotic rules. Still on the contrary, living organisms do not have a changeable relation to natural laws but *are subjugated* in a strict manner, e.g. gravity.

Hence, rules of sign use may be altered, rearranged, invented, generated or delated, i.e., communication processes can function but in every case they may also fail, i.e., there is no natural law which garantees their success. The following of semiotic rules in sign-use and interpretation is different to .e.g. subjugation under natural laws as it happens in the cristallization of water to ice. This is the case also with the genetic code. If the code functions as code this is physically and chemically coherent. The biotic plus is that this code functions additionally also syntactically, pragmatically and semantically in parallel, whereas the freezing of water to ice does not. There is no semiotic interpretation of water molecules which are exposed to some degrees under zero but only physical and chemical laws.

Each of these levels of semiotic rules may be damaged or deformed by several influences and circumstances, so that the content of the code is not correct, i.e., is not able to serve as matrix for appropriate protein production (coding) or as element of higher order regulation (non-coding). This difference marks the limitation between non-living matter and living nature.

Therefore it becomes clear that the conditions on the surface of a planet without living organisms are determined entirely by the sun's energy and the laws of physics and chemistry, as James Lovelock showed convincingly. A planet with living organisms deviates considerably from this scheme. The gas composition and temperature will vary in a manner that cannot be predicted exclusively by the laws of physics and chemistry alone. Such a planet contains incompatible gas mixtures and temperatures whose relatively stable balance is actively controlled by organisms. And these organisms are no solus ipse subjects but a part of a planetary symbiology which is organized and coordinated by biocommunication.

Biosemiotics defines a biotic "plus"

If we choose the communicative approach we can culminate towards a new transparent and for every living organism coherent method of description which pertains to the levels of interaction in which the organism is integrated. The living world is not outside to the outstanding humans, but we are all part of living nature with indispensable interrelations. The old category system that takes humans on one side, the environmental living beings on the other is not appropriate nowadays. Biology is a priori symbiology (see Chapter 6). Without symbioses there would not have been something like living nature in principle. And beneath intraspecific and species-specific sign use there is a level of trans-specific sign use also.

It is clear that the "biotic plus" of living nature and therefore the difference between non-living and living nature started with a very few or the only event (see quote on bookcover backside). Living nature is constituted when semiotic use of chemical molecules starts, in parallel; it is the initial stage of raising up semiotic complexity. Living nature begins with the use of sign(al)s between entities which are in their turn able to use, produce and interpret signs under syntactic, pragmatic and semantic rules. This process could also fail and turn back in physico-chemical realm without semiotic rules again. Starting of living nature from biosemiotic perspective may have had several trials and errors.

The advantages of biosemiotic symbiology

So the aim of a biosemiotic symbiology is to observe and describe the semiotic rules of sign use in living nature. If the description is correct our understanding of coordination and organisation of evolution, development, growth and disease in the living nature (cells, tissues, organs, organisms) will expand. This could be an advantage in medical sciences as proven by the work of Thure von Uexküll.

In some cases we will be able to predict vectors of development. But the most important advantage of this new type of biology (as a kind of understanding social science) over mechanistic biology is that we can be aware of the complexity of living nature to prevent careless or even negligent steps in technical developments, i.e., to use artificial bio-engineering without being aware of possible consequences. It is a crucial difference whether we try to handle a mechanical functioning matter or a communicatively organized living nature. In the first case we can trust laws of physics and chemistry exclusively; in the second we must be aware, that change and reorganisation, success and failure of biotic agents are common and an open processs in principle.

Linguistic and communicative competences in non-human nature

The three semiotic levels of rules are necessary to generate a common usable repertoire for sign-mediated interactions, i.e. communication processes have a complementary function. This means that if one level is reduced to another (or to the remaining others) no adequate analysis of sign use is possible: through analysis of syntax or semantics alone it is not possible to understand or even reconstruct the pragmatic interactional content of two communicating indviduals, and vice versa. This is a crucial difference between a three-leveled biosemiotics and e.g.

bioinformatics with its main focus on genetic syntax or biosemantics with its accent on semantic functions.

There is also a difference to (A) construct correct sentences or genetic sequences according to syntactic, pragmatic and semantic rules or (B) to generate an appropriate behavior in which the sequence of behaviours, e.g. attack, defense, mating, virulence, common coordination for a special purpopse takes the function of signs. The language-like competence to invent or rearrange the genomic content is a kind of *linguistic competence* which includes other semiotic rules than the *communicative competence* which initiates and constitutes a context-coherent interaction. So the biosemiotics of genome editing includes other issues than biosemiotics of e.g. plant communication. Although these two levels of biosemiotic processes are connected in natural genetic and even more in natural epigenetic engineering.

Non-reducable complementary roles of linguistic competences and living organisms

As we will see in Chapter 5, the linguistic competences of viruses (and especially their competence to constitute the sequence ("word") order in bacteria) leads to all necessary prerequistites for the genetic and genomic as well as proteomic evolution. But the linguistic competence needs living organisms which are different, needs a biotic "matrix" to expand these competences. Without living and interacting bodies, cells, etc. the genomic creativity would be only a possibility restricted to mere RNA combinatorial events without being relevant to generate a biosphere. Agents with *genome editing competences* need a real lifeworld of interacting living bodies to get different situations as matrix for inventions or genomic creativity, as do organisms which cannot live without continued natural genetic engineering and genome editing within their cells. This means linguistic competences need the complementary role of communicative competences.

New view on living nature opens new qualities for future decisions

The origin of life is not question of this book which is focused primarely on communication processes, i.e., rule-governed sign-mediated interactions within and between living organisms on all levels in all organismic kingdoms. Furthermore, it develops theses for an appropriate description of the linguistic competences of agents which are editing the genetic/genomic contents.

It should provide a better understanding of living nature in general and also of us. In difference to pure knowledge shuffling for commercial purposes this biosemiotic approach may help to change the relationship of human beings to non-human living nature. This non-human nature is not the environment, the *Umwelt*, but in a symbiotic living nature a *Mitwelt*, a co-world (Chapter 7) which we depend on vitally. If our ecological "footprints" are further on destructive we contradict symbiological rules of living nature. This could mark a kopernican turn in our relation and understanding of life in general.

It is our decision which pragmatic directions we want to practise, a destructive, parasitic relation of exploitation, pollution and extinct of species or a responsible sustainable and symbiotic-like cooperation with living nature to human and non-human benefits. The success of the latter one will decide in the long run, whether our self-understanding of the top of evolution was correct or in opposite human mankind was an evolutionary error construction. As living organisms we have to decide (not as solus ipse subjects but by major votes), to communicate to coordinate and to organize such practical behaviour either in a sustainable coherence to living nature or as overdeveloped hybrids of the species anthropos.

1

Plant Communication from Biosemiotic Perspective

Differences in abiotic and biotic signal perception determine content arrangement of response behavior. Context determines meaning in meta-, inter- and intraorganismic plant signaling[*]

Abstract: As in all organisms, the evolution, development and growth of plants depends on the success of complex communication processes. These communication processes are primarily sign-mediated interactions and not simply an exchange of information. They involve active coordination and active organization – conveyed by signs. A wide range of chemical substances and physical influences serve as signs. Different abiotic or biotic influences require different behaviors. Depending on the behavior, the core set of signs common to species, families, genera and organismic kingdoms is variously produced, combined and transported. This allows entirely different communication processes to be carried out with the same types of chemical molecules. Almost without exception, plant communication are parallel processes on multiple levels, including (A) between plants and microorganisms, fungi, insects and other animals, (B) between different plant species and between

[*] Plant Signaling & Behavior 1(4) 2006.

members of the same plant species (C), as well as between cells and in cells of the plant organism.

Keywords: multilevel plant-communication, sign-mediated interactions, context-dependency

1. Introduction

Based on their apparently static life form, plants have been traditionally been viewed and treated as growth automatons. Today, however, we recognize that the coordination of growth and development in plants, like in all other organismic kingdoms, is possible only by using signs (Greek: semeion) rather than pure mechanics. Understanding the use of signs in communication processes requires a differentiated perspective. Chemical molecules are used as signs. They function as signals, messenger substances, information carriers and memory medium in either solid, liquid or gaseous form.

In this review I will demonstrate that plants are sessile, highly sensitive organisms that actively compete for environmental resources both above and below the ground. They assess their surroundings, estimate how much energy they need for particular goals, and then realize the optimum variant. They take measures to control certain environmental resources. They perceive themselves and can distinguish between self and non-self. This capability allows them to protect their territory. They process and evaluate information and then modify their behavior accordingly.

To understand this highly diverse competences we will notice, that this is possible due to parallel communication processes in the plant body (intraorganismic), between the same and different species (interorganismic), and between plants and non-plant organisms (metaorganismic). Successful communication processes allow the plants to prosper, unsuccessful ones have negative, potentially lethal repercussions. Intraorganismic communication involves sign-mediated interactions in cells (intracellular) and between cells (intercellular). Intercellular communication processes are crucial in coordinating growth and development, shape and dynamics. Such communication must function on both the local level as well as between widely separated plant parts. This allows plants to react in a differentiated manner to its current developmental status and physiological influences.

As we will see, communicative competence refers to chemical and physical communication processes. Chemical communication is either vesicular trafficking or cell-cell communication via the plasmodesmata. Moreover, numerous signal molecules are produced in or controlled by the cell walls. Physical communication takes place through electrical, hydraulic and mechanical signs.

It should be noted that signs, whether abiotic or biotic, are interpreted. This means they must be identified as components of messages that differ from molecules that are not components of messages (noise). The interpreter is always a living individual. The interpretations can either be successful or unsuccessful. Thus, the message is perceived in its correct sense and meaning and a tailored response behavior is generated. Or it is misinterpreted – sense and meaning are perceived in a distorted or deformed manner – and the response behavior fails to appear or is inappropriate.

We will recognize that the use of molecular languages/codes goes beyond information exchange: it produces various active behaviors and interactions. The many types of symbiosis show that behavior towards the symbionts can be mutually benefit and stress-free. Such relationships can change when this balance is lost, for example when one partner is weakened. The interaction level shifts and one partner views the other as a source of stress. In plants, altruistic forms of interactions occur even in the root zone, as do life-and-death defensive battles. In every case, the situational context determines the meaning of the signs.

We will see that sign-mediated interactions within and between organisms are possible due to the fact that living individuals share a core set of signal molecules with members of their own species but also with members of other species, families, genera or organismic kingdoms: these molecules are produced and emitted at specific levels, amounts and rhythms. The relationship between the molecules is governed by specific rules. The molecular syntax [7,8] of molecular languages/codes determines the correct sequence and combination of signal molecules. Disrupting or deforming these syntactic rules can cause incomplete transmission of the message, triggering faulty interpretations and responses in the receiver.

A completely different set of rules determines the interaction behavior between organisms, cells and tissues: growth and development are other forms of behavior than defense or sexual reproduction; mutualistic symbioses require pathways that differ from those in commensalism or parasitism. One and the same core set of species-specific signal molecules are used in different interactions to produce different pathways. Moreover, one and the same pathway can take on different meanings (semantic functions) in different interactions and trigger different responses by one and the same receiver. A purely syntactic or semantic analysis cannot explain this because it cannot identify the pragmatic rules that determine the concrete interactions. This calls for considering all three levels of semiotic rules, as in any other analysis of sign use in living nature, and it will show the full range of multilevel communicative competences of plants.

2. Chemical vocabulary

The chemical communication in and between plants is so complex that more than 20 different groups of molecules with communicatory function have currently been identified. Up to 100,000 different substances, known as secondary metabolites, are active in the root zone, for example. This diversity is necessary considering the high diversity of microbes, insects and plants in this zone.[9] For example, the continuous defense against pathogenic microorganisms in the root zone requires the constant production, exact dosage and secretion of phytoalexins, defense proteins, and other substances.[10] Here, I present selected examples of the molecular vocabulary in plant communication:

Contextdependent Auxin as neurotransmitter, hormone, morphogenic sign

Plant roots and plant shoots detect environmental signals as well as development levels and communicate over long-distance pathways. The decentralized nervous system of plants is advantageous for decentral growth and development under constantly changing environmental conditions.[11] Auxin is used in hormonal, morphogenic and transmitter pathways. Because the context of use can be very complex and highly diverse, identifying the momentary usage is extremely difficult.[12] For synaptic neuronal-like cell-cell communication, plants use neurotransmitter-like auxin[13] and presumably also neurotransmitters such as glutamate, glycine, histamine, acetylcholine, dopamine – all of which they also produce.[11] Auxin is detected as an extracellular signal at the plant synapse[12] in order to react to light and gravity. However, it also serves as an extracellular messenger substance to send electrical signals and functions as a synchronization signal for cell division.[14] In intracellular signaling, auxin serves in organogenesis, cell development and differentiation. In the organogenesis of roots, for example, auxin enables cells to determine their position and their identity.[15] The cell wall and the organelles it contains help regulate the signal molecules. Auxin is – as the name suggests – a growth hormone. Intracellularly, it mediates in cell division and cell elongation. At the intercellular, whole plant level, it supports cell division in the cambium, and at the tissue level it promotes the maturation of vascular tissue during embryonic development, organ growth as well as tropic responses and apical dominance.[16]

Hormones

Alongside the classical phytohormones auxin, cytokinin, gibberellin, ethylene and abscisic acid, the plant peptide hormone systemin has noticed to be important; plants use this to systematically react to local injuries.[17]

For example, the abiotic stress hormone absisic acid imparts disease resistance by acting on several levels involved in biotic stress signaling.[18] Peptide signal-mediated responses are merely one part of a biological process that is controlled by a combination of several hormones. In activating an effective defense response, a combination of systemin, jasmonate and ethylene serves as signal molecules.[17]

The production (biosynthesis) of brassinolide hormones is important for cellular processes and development steps. They are therefore termed metahormones.[19] Arabidopsis plants that lack this hormone remain small and are male-sterile. Many plant hormones apparently play a key role as signals in cell functions and developments that enormously impact the activities of insects. Plant hormones control not only plant growth and development but also serve in communication within the same species, with related or unrelated plant species, and with insects, i.e. they serve in classical metaorganismic communication. The fact that plants and insects produce their hormones differently but apply them for similar purposes, namely to coordinate overall development, points to their use in their unicellular ancestors.[20]

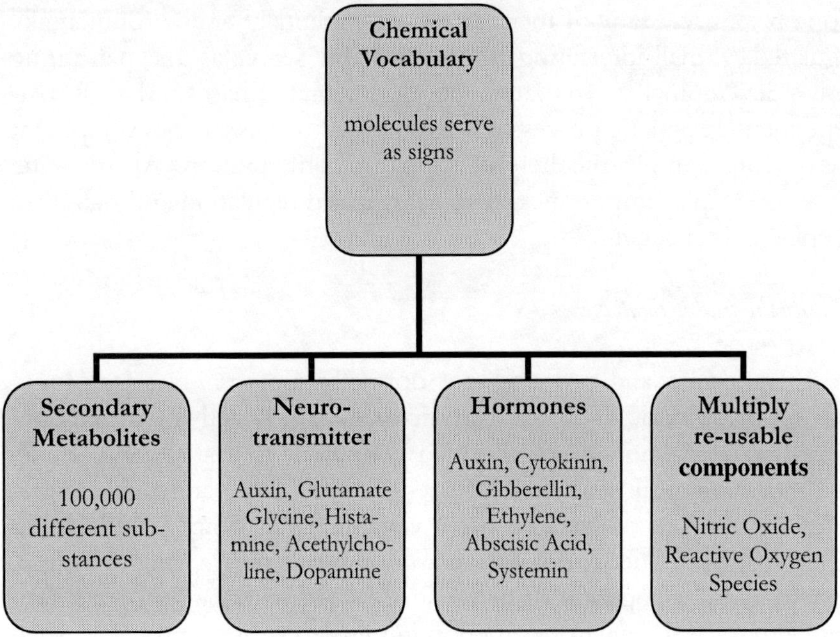

Figure 1: Examples for chemical vocabulary in plant communication

RNAs

Sessile organisms can react to the full range of outside influences only through behaviors that are expressed in growth and development; correct timing, which can be very precise, is crucial.[21] Beyond phytohormones, the chemical messenger substances include peptides such as phytosulphokine growth factors and RNAs. Micro-RNAs play an important role in intracellular communication during plant development, either in cleavage during translation/transcription or in preventing translation. Micro-RNAs are apparently necessary for meristem function, organ polarity, vascular development, floral patterning and hormone response. Many of them are developmentally or environmentally regulated.[22] Small interfering RNAs probably serves as a signal during early development. In later developmental phases, the RNAi-dependent epigenetic processes are reminded of this early development phase, for example the heterochromatin configuration. At any rate, these RNAs play important roles in chromatin regulation and therefore in epigenetic silencing.[22]

Multiply re-usable components

Small molecules and proteins that normally support important functions in plant immunity, such as nitric oxide and Reactive Oxygen Species (ROS), have now been identified as multiply re-usable components of other biological processes. Messenger substances and signal molecules are used as a versatile basic vocabulary in other contexts and other regulation networks – a common principle in the evolution, growth and development of organisms.[23,24] Nitric oxide (NO) is a substance that has a regulatory function in numerous signal processes such as germination, growth, reproduction and disease resistance.[25] The same is true for diverse species of ROS.[26, 27]

3. Interpretation of mechanical influences

Mechanical contact has an influence on the overall organism and on the cell level, both in plants and in other eukaryotes. Contact can cause plants (A) to react aggressively, for example toward the animals that want to eat them, (B) to discard their pollen, and (C) can cause the plant stem to grow into the sunlight.[28] The entire configuration of a plant (morphogenesis) is partially determined by mechanical inputs, for example wind and gravity (gravisensing[29]). Responses to contact involve signal molecules and hormones along with intracellular calcium, reactive oxygen, octadecanoids and ethylenes. Another common feature is contact-related gene expression. Many of these genes code for calcium bonds, cell wall changes, defense, transcription factors and kinase proteins.[28]

The detection of resources and their periodic, cyclic availability plays a key role in plant memory, planning, growth and development. When, for example, young trees obtain water only once a year, they learn to adjust to this over the following years and concentrate their entire growth and development precisely in the expected period.[30]

Interpretation processes in the plant body are highly sensitive. In taller-growing plants, for example, the water balance places enormous demands on cell wall development and cell wall structures, which must adapt to the (often extreme) pressures involved in storage and pressure distribution. A sophisticated and multi-leveled feedback- and feedforward-system guarantees a plant-compatible water balance even under extreme environmental conditions.[31,32] To date, 7 different levels of sensitivity to water shortage have been described. They are based on the different types of physiological and phenotypic responses.[33] Plants are especially sensitive to light and have various receptors for UV, blue, green, red and far-red light.[33] The angle of the light, combined with sensation of the growth of adjoining plants, is decisive in enabling plants to coordinate their growth with respect to the optimal light angle and shade avoidance.[34] The adaptive response of the plant, i.e. altered growth, depends on the seconds-, minutes- and hours-long dominating wavelength of the incoming light, and on the combination of wavelengths across the whole day. The roots receive constant signals from the aboveground parts of the plant for specific growth orientations.[35]

4. Metaorganismic (transspecific) communication

Sign-mediated interactions with organisms belonging to other species, genera, families and organismic kingdoms are vital for plants and are coordinated and organized in parallel. They are almost always symbiotic or parasitic and range from mutually beneficial via neutral, up to damaging behaviors. The different forms of symbiotic communication require very different behaviors from the participating partners. This involves large numbers of complementary direct and indirect defense behaviors.

Coordination of defense against pests and injury

A good example of parallel meta-, inter- and intraorganismic communication are coordinated defense strategies of plants. Chemical signal substances are the oldest form of signs and are used by microbes, fungi, animals and plants. They are transmitted via liquids in the environment or within the plant body; they can be distributed and perceived through the atmosphere. Leaves always emit such volatiles in small doses, but emit greater quantities when infested by parasitic insects. This allows them to attack the parasites either directly by producing substances that deter them, or indirectly by attracting other insects that are natural enemies of the parasites. These volatiles are also perceived by neighboring plants, allowing them to initiate preemptive defensive responses.[36] Volatile phytochemicals serve as airborne *semiochemicals*. Depending on the behavioral context – destruction, injury or parasitic infestation – the emitted scents clearly differ for both the insects and neighboring plants.[36] The plants coordinate complementary direct and indirect defense mechanisms in a step-wise manner and tailor them flexibly to the severity of the injury or the density of pest infestation.[37,38]

When plants are attacked by pests, they develop immune substances that function the same as in animals.[39] Injured plants produce aromatic substances that warn other plants. They then rapidly produce enzymes that make the leaves unpalatable for herbivorous insects. Rather than being passive prisoners of their surroundings, plants are active organisms[40] that identify their pests and actively promote the enemies of these pests.[41]

In lima beans, for example, a total of 5 different defense strategies against mite infestation have been discovered. First, they change their scent to make them unattractive to the mites. Then the plants emit scents that are perceived by other plants, which then do precisely the same thing to warn surrounding lima beans before the mites even reach them. Some of the emitted substances had the effect of attracting other mites that ate the attacking red mites.[42] Similar defense processes have been described in tomato plants.[37,43]

Plants possess a non-self warning system to fend off dangerous parasites. So-called pattern recognition receptors detect patterns of chemical substances associated with parasite infestation.[44] The microbes, in turn, react to this pattern recognition.[45]

Because plants are sessile, their reaction potential is geared toward defense against mechanical damage and pest infestation. One of the many reaction types to infestation is the production of protease inhibitors I und II, which block protein degradation in the digestive tracts of insects. This defense reaction is produced both at the injured site and throughout the surrounding tissue: the local wound response triggers the production of mobile signals that prompt a systematic reaction of the overall plant.[17]

Plant roots have the capacity to produce 100,000 different compounds, largely secondary metabolites, many with cytotoxic properties, in order to prevent the spread of microbes, insects and other plants.[9,46] For example, plants have developed defense strategies in which substances are emitted in the root zone such as signal mimics, signal blockers and/or signal-degrading enzymes to respond to bacterial quorum sensing.[46] In the defensive position, they can disrupt the communication of parasitic microorganisms to the point that the internal coordination of the parasitic behavior collapses.

"Friendly" arthropods such as predaceous or fungivorous mites are supported by plant "domatia", similar to the situation in complex communities of grasses and fungal endophytes. These symbiospheres, however, can also be misused, for example by mites that colonize these domatia for themselves without benefiting the host cell.[47]

Communicative coordination of symbioses

A limited number of chemical messenger substances is available to maintain and simultaneously conduct the communication between (A) root cells of three different types, (B) root cells and microorganisms, (C) root cells and fungi, and (D) root cells and insects.[9,46,48-51] The communication process in the root zone is generally meta-, inter- and intraorganismic and requires a high communicative competence in order to be successfully interactive on all three levels and to distinguish messenger molecules from molecules not being part of messages.[52-54]

It has been postulated that the origin of root cells in plants, and therefore the basis for the youngest organismic kingdom on our planet, arose through the symbiogenesis of fungi and algae.[35,55,56] One hypothesis assumes that land plants are the symbiogenetic product of green algae and a tip-growing fungus-like organism that combined autotrophic and heterotrophic capabilities.[57]

Vital Symbiosis of plant roots with bacteria, fungi and insects

Plants use their plant-specific synapses[12] to conduct neuronal-like activities and establish symbiotic relationships with bacteria.[58] Similar mutually advantageous relationships are established with mycorrhizal fungi.[59] A special type of plant synapse resembles the immunological synapse of animal cells and allows plants to respond to pathogen and parasite attacks as well as to establish stable symbiotic interactions with rhizobia bacteria and fungal mycorrhiza.[12,60-64] Electrical signals can reinforce chemical signals or overcome short-distance responses of fungal mycelia that can be present on root surfaces.[65] Interestingly, rhizobia bacteria are taken up in plant cells via phagocytosis during symbiotic interactions with roots of leguminous plants.[66]

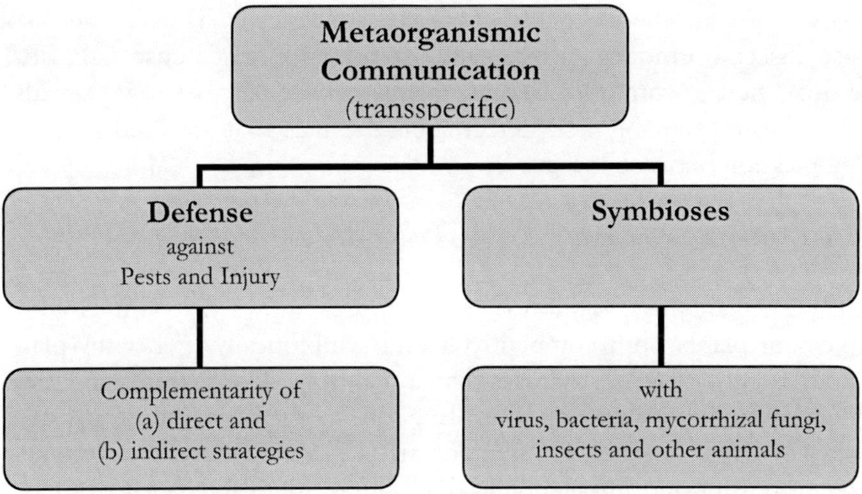

Figure 2: Examples for metaorganismic (transspecific) communication

The symbiotic relationship between legumes and rhizobial bacteria leads to the formation of nitrogen-binding nodules in the root zone. Nod factor signaling and thigmotrophic responses of root hairs overlap here as well. This once again shows how the same pathways are used for different signal processes.[67]

Today, several hundred species of fungi colonize more than 100,000 different plant species. This type of cohabitation requires symbiotic signaling.[68] Roots develop from rhizomes in order to provide better conditions for mycorrhizal fungi, which in turn supply plants with better nutrients.[69] For the fungus the relationship is either balanced or predatory. Endophytic fungi, however, live in plants without triggering disease symptoms.[69] Similar to the symbiosis between plants and mycorrhizal fungi, the symbiosis between asexual endophytes and grasses also represents a type of complementary parasitism.[70]

Plants, insects and microbes share a particular repertoire of signals. Some are therefore also employed strategically. Thus, plants also use insect hormones (prostaglandins) for specific defense behavior. Signal theft is common. Because plants can detect their own signals, they can presumably also detect similar signals that are used in communication between insects.[71]

Viral symbiotic interactions

In particular, the evolution of plant viruses shows that viruses complement plants both competitively and symbiotically. A healthy plant body is better for most viruses than a sick body. Plant viruses and their development provide a good explanation for the observation that new species originate through symbiogenesis.[72] Viruses use intergenomic gene transfer and intragenomic duplication. Many DNA viruses have encoded numerous nucleic acid metabolisms that are very similar to cell proteins. Examples include DNA polymerases, ribonucleotide reductase subunits, DNA-dependent RNA polymerase II subunits, DNA topoisomerase II, thymidylate synthase, helicases and exorbinuclease. Viruses probably invented DNA to protect their genetic material from being changed by RNA or RNA encoded enzymes.[73]

One of the interaction processes between plant viruses and their host organisms creates a defense level against foreign nucleic acids.[74]

Plant viruses code for silencing suppressors in order to act against host RNA silencing, and some of these suppressors effect micro-RNA multiplication and hinder plant development.[75] But also viroids play a symbiotic role. Despite their small size and their non-coded genome, viroids can multiply, systematically spread from cell to cell, and trigger symptoms in the host.[76]

5. Interorganismic communication

Research has shown that plants can distinguish between damage caused by insects and mechanical injuries. Mechanically injured plants emitted substances that were ignored by neighboring plants, whereas they all reacted immediately to pest infestation.

Plants can distinguish between self and non-self. Thus, defense activities are initiated against foreign roots in order to protect the plant's own root zone against intruders. The individual sphere of a root, along with its symbiotic partners, requires certain fundamental conditions in order to survive and thrive. When these prerequisites are threatened by the roots of other plants, substances are produced and released in the root zone that hinder this advance.[9,46,49,50] Such defense activities are also deployed as anti-microbial substances against the microflora in the root zone.

Plant roots produce a wide range of chemical substances: (A) some enable species-specific interactions; (B) many of these substances are released tens of centimeters into the surroundings; (C) these substances have strong but not necessarily negative effects on animals, bacteria, viruses and fungi; (D) released substances have a defensive function against other plants; (E) many substances have absorptive characteristics that reduce the negative effects of substances.[9]

As reported above in lima beans and tomatoes, also corn plants use a sophisticated communication system to warn each other about pests. By emitting green leafy volatiles, the corn plants attract the natural enemies of the pests and alarm neighboring plants. The alarmed neighbor then produces a protective acid that is normally produced only in response to external injuries.[38]

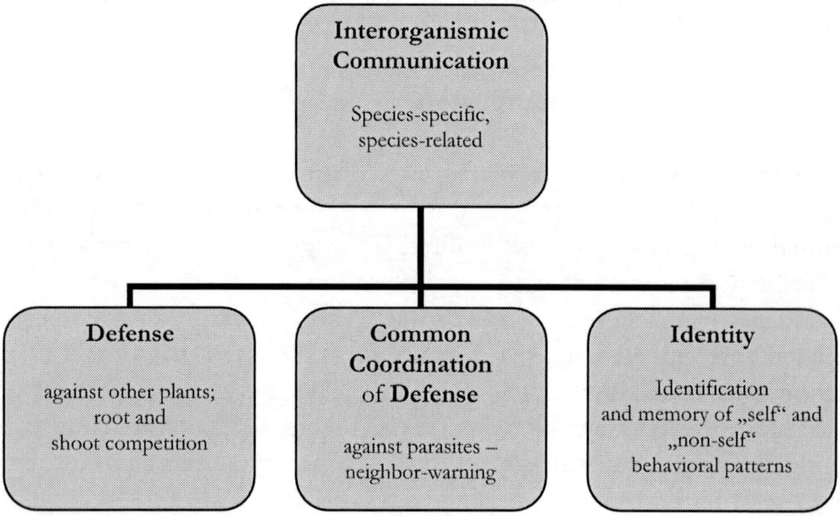

Figure 3: Examples for interorganismic (species-specific, species-related) communication

Plants use biotic signals to inform each other about the presence, absence and identity of neighboring plants, growth space, growth disturbances and competition.[48] Plants that are removed and planted elsewhere remember the identity of their former closest neighbors for several months.[77] Recognition patterns in neuronal-like networks are one possible explanation.

Parasitic plants are an important feature in the plant world. Today, about 4000 species have been described. In order to parasite other plants, their root apices transform into fungal-like haustoria which extract photosynthates from vascular tissue of pray roots.[61,78] Parasitic plants are present wherever other plants can grow, from the tropical rainforest to the Arctic, and take important nutrients and environmental resources (light) away from non-parasitic plants. They therefore influence entire ecosystems, population dynamics, and biodiversity, including the presence and diversity of microbes, birds, insects and other animals.[79]

6. Intraorganismic communication

As opposed to the central nervous system of animals, which controls metabolism and reactions centrally, the control in plants is decentral.[80] This enables plants to start independent growing or developmental activities in certain regions of their body, for example on how a particular branch should grow, depending on the wind, light angle and overall "architecture" of the plant.[33] Most of the activities that plants make with regard to growth and development require communication processes – synapse-like communication – between all parts of the plant.

Intercellular communication

Short-distance communication differs considerably from long-distance communication. As a rule, both complement each other. Intercellular communication in the root zone (in the soil) differs from that in the stem region above ground. Both are necessarily coordinated with one another in order to enable life in these different habitats. Intercellular communication informs other plant parts about events in specific organs or regions of the plant (especially in large plants), for example sugar production in leaves, the reproduction in flowers and resource utilization by the roots.[81]

Plant cells are connected by plasmodesmata. These connecting channels enable the flow of small molecules as well as ions, metabolites and hormones, and allow the selective exchange (size exclusion limit) of macromolecules such as proteins, RNAs and even cell bodies.[82] The plasmodesmata impart plants with a cytoplasmatic continuum known as the symplasm.[76] But plasmodesmata are more than mere transport channels; they also regulate and control the exchange of messenger substances in a very complex manner.[83] In symplastic signaling, the intercellular communication of plants differs fundamentally from that in other organismic kingdoms.[84] It integrates various communication types such as local and long-distance communication. Beyond symplastic communication (especially in the meristem, where new tissues are produced), plants also exhibit the receptor-ligand communication typical of animals.[84] While receptor-ligand communication determines stomatal patterning in the epidermis of mature leaves, trichome patterning is mediated by symplastic signaling.[85]

For long-distance signaling movement proteins play an important role. Movement proteins convey information bearing RNA from the stem and leaves to the remote roots and flowers. The movement protein allows the mRNA to enter the plasmodesmata tunnel, into the phloem flow. Once it has entered this transport system, it can relatively rapidly reach all parts of the plant. These RNAs can control the levels of other proteins. The level contains information for local tissues, for example about the general physical condition of the plant, the season, or the presence of dangerous enemies.[81]

Plasmodesmata are prerequisites for intercellular communication in higher plants.[86] In embryogenesis they are an important information channel between fetal and maternal tissue. The further the development of the embryo, the more reduced the cell-cell communication between embryo and maternal tissue.[87] Cell-cell communication via direct transmission of transcription factors plays a central role in root radial and epidermal cell patterning as well as in shoot organogenesis.[88]

The cellular organization of the roots is determined during the plant's embryonic development and is controlled by intercellular communication. Bonke et al.[89] provide a particularly good example of communicative control of these 10 phases of embryogenesis. This confirms the presence of local signaling centers and the complex relationship between numerous different signaling pathways.

A wounded plant organizes an integrated molecular, biochemical and cell biological response. This strategy enables information to be transported across great distances, for example in tall trees.[90] Proteins that can be detected by receptors enable a thoughtful response[91] by plants. There are about one thousand known protein kinases/phosphatases, numerous secondary messengers and many thousands of other proteins.[33] Through their life cycles and their growth zones, plants develop a life history of environmental experience that they can pass on to later generations and, should they themselves grow to be several hundred years old, utilize themselves.[33] Even small plants store stress experiences in their memories and then use these memories to coordinate future activities.[92] Especially during growth, key information about the current status often takes a back seat to future-oriented processes, for example early root growth and nutrient supply to secure future developments such as larger leaves. From this perspective, plants must plan for the future and coordinate growth, food uptake and communication with symbionts.[93]

The complementary differentiation of communication types into short-distance and long-distance signaling – with their different yet ultimately complementary tasks – requires cells to identify their position. They accomplish this by, among other things, detecting signals from neighboring cells.[94] Thus, the identification competence of "self" and "non-self" by cells can be interpreted as a result of social interaction rather than solipsistic behavior. For example, signals from leaves trigger flower development at the tip of a plant.[94] An entire network involving 4 different signal pathways regulates this transition from the vegetative to the reproductive phase.[94] Most flowers bear closely adjoining male and female reproductive organs. Self-incompatibility is therefore crucial in distinguishing between own (related) and foreign (non-related) pollen. This self-/non-self differentiation ability is promoted by signal processes also used in other plant responses.[95]

Signals amend one another to form signal sequences, much like words combine to form sentences: different active forms of behavior determine the combination and production of molecule-sentences. This distinguishes cell differentiation during root development from cell differentiation during stem development, or developmental processes during the vegetative phase from developmental processes in the reproductive phase.

Intracellular communication

Intracellular communication in plants takes place between the symbiogenetically assimilated unicellular ancestors of the eukaryotic cell, mainly between the cell body and cell periphery. It transforms and transmits external messages into internal messages that exert a direct (epigenetic) influence on the DNA storage medium and trigger genetic processes; this leads to the production of signal molecules that generate a response behavior. Via endocytosis, however, bacteria, viruses and viroids interfere with this intracellular communication and can support, disrupt or even destroy it. Intracellular communication offers viruses the opportunity to integrate certain genetically coded abilities of the host into their own genome or to integrate their own genetic datasets into the host genome. The ability of viruses to integrate different genetic datasets probably plays a major role in symbiogenetic processes.

The eukaryotic cell is composed of a multicompetent nucleus as a basic building block of life and a cell periphery-"apparatus" that was symbiogenetically the ancestor of other endosymbionts. Interestingly, both the nucleus and viruses have several similar features and capabilities: they both lack the protein synthesis "machinery" and the fatty acid-producing pathways. Both transcribe DNA but do not translate it into RNA. Viruses were probably very important in the evolution of eukaryotic cells because they were able to conduct cell-cell "fusion".[96] There are strong reasons too, that the eukaryotic nucleus is of viral origin.[97-99]

Neuronal plasticity refers to the ability of neuron populations to alter – to either strengthen or weaken – their connections *based on experience.* This is the basis for learning and memory. Like memory, long-term neuronal plasticity requires new RNA and protein synthesis. Accordingly, the signals must be transported from the synapse, from where they are sent, to the nucleus, where they are transformed to change the gene transcription. Then, the products of gene transcription (proteins, RNAs) must be sent back to the synapse in order to permanently change synaptic strength. This communication process is well described in animals;[100-102] if plants exhibit neuronal plasticity, then similar descriptions may follow.

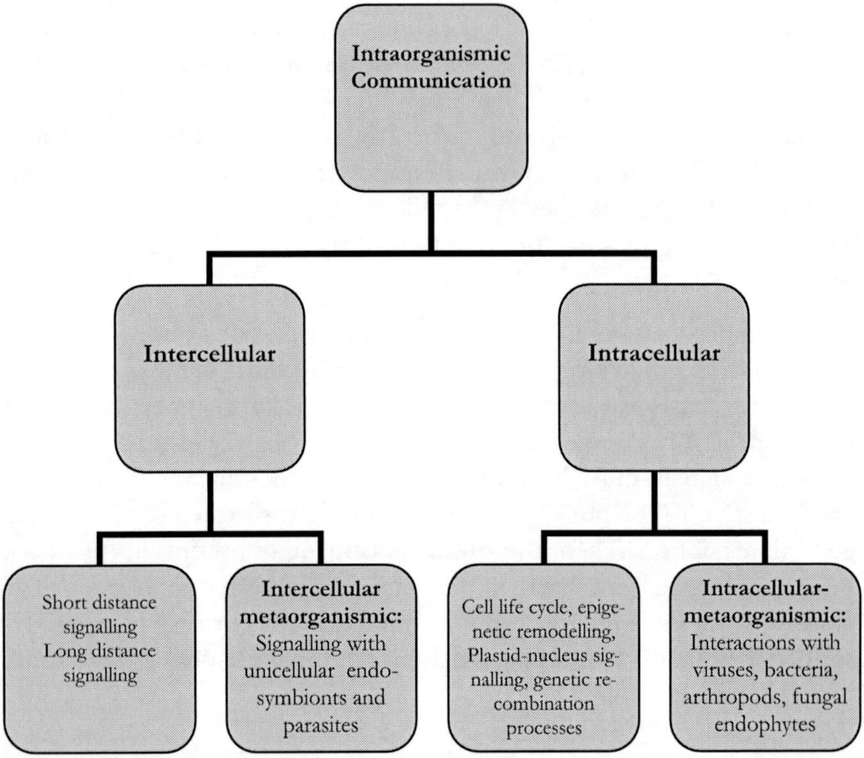

Figure 4: Selected examples for intraorganismic communication

Reports on the transfer of mitochondrial genes between unrelated plant species caused some surprise. While gene transfer is an extremely rare event in animals and fungi, it is common between plant mitochondria.[103] Variations in repetitive DNA that manifest themselves as variation in the nuclear DNA complex have far-reaching ecological and life history consequences for plants.[104]

The function of a eukaryotic cell depends on successful communication between its various parts. Plastids send signals to regulate nuclear gene expression and thus to reorganize macromolecules in response to environmental influences.[105] It has been shown that micro-RNAs regulate certain developmental processes such as organ separation, polarity and identity, and that they define their own biogenesis and function.[106] Eukaryotic genomes are regionally divided into transcriptionally active euchromatin and transcriptionally inactive heterochromatin.[107] Epigenetic changes can also take place without changes in genomes, for example through various inactivations and activations of genetic datasets via chromatin remodeling, transposon/retro release, DNA methylation, novel transcription, histone modification, and transcription factor interactions.[108] Epigenetic changes are also reversible.[109] Various stress situations in plants are known to cause transposon movements,[110] and bacterial infections or UV stress can cause chromosomal rearrangements,[111] i.e. changes in higher-order regulation levels that control the transcription processes of the protein-coding DNA.

Repetitive DNA is present in two syntactic combinations: tandem repeats and dispersed repeats. Tandem repeats consist of sequences that can contain several thousand copies of elements that are dispersed throughout the genome. Pericentromeric sequences consist of a central repetitive nucleus flanked by moderately repetitive DNA. Telomeric and subtelomeric sequences consist of tandem repeats at the physical end of the chromosomes. Retroelements and transposable elements are involved in replication and reinsertion at various sites in complex processes: these include activation of excision, DNA-dependent RNA transcription, translation of RNA into functioning proteins, RNA-dependent DNA synthesis (reverse transcription) and reintegration of newly produced retroelement copies into the genome.[104]

Endocytosis and vesicle recycling via secretory endosomes are indispensable for many processes in multicellular organisms. Plant endocytosis and endosomes are important for auxin-mediated cell-cell communication as well as for gravitropic responses, stomatal movements, cytokinesis and cell wall morphogenesis. As in animals, synaptic cell-cell communication is based on rapid endocytosis and vesicular recycling in plants.[112]

Plants can overwrite the genetic code they inherited from their parents and revert to that of their grand or great-grandparents.[113-115] This contradicts traditional DNA-textbook conviction that children simply receive combinations of the genes carried by their parents. Now a backup code has been found; it can bypass unhealthy sequences inherited from the parents and revert to the healthier sequences borne by their grandparents or great-grandparents. Research has shown that plants are able to replace abnormal parental code sequences with the regular code possessed by earlier generations. Does this require inheritance not only of the parental genetic make-up but also that of the grand-parents and former ancestors? What is proposed is that higher-order regulation function in non-coding DNA – a type of genome-editing MetaCode [116] – save ancestor genome structures, which overrule protein-coding DNA under certain circumstances like stress. This means that the (pragmatic) situational context of the living plant body may induce epigenetic intervention on the genome editing MetaCode, i.e. active micro-RNAs activate a certain signaling pathway network which can restructure the semantics of a genetic make-up. By initiating chromosomal methylation and histone-modifications, certain silencings, start and stops, and alternative splicing processes constitute alternative sequences. The result is that, in the existing genome architecture, not the inherited parental sequences are translated and transcribed but the backup copy of grand or great- grandparents. Under normal conditions, the operative genetic make-up stems from the parents. These research results indicate that not only is a combination of parental genes inherited, but also ancestral genome-regulating features in "non-coding" DNA; this enables alternative splicing pathways, i.e. a different use and multiple protein meanings of one and the same genetic data set.[113-115]

7. Outlook

Plants are the youngest organismic kingdom and perhaps the main success story of evolution. They arose ca. 350 million years ago, and terrestrial plants, which flower and bear fruits (a key prerequisite for feeding in larger animals), only developed 150 million years ago. Higher plants make up 99% of the biomass on our planet; of this, nearly 84 % are trees. The lack of mobility is often construed as a disadvantage vis-à-vis representatives of the animal kingdom. From an objective perspective, such immobility and the sessile life style must have been an advantage. Plants are clearly the most malleable of organisms, a trait that can be attributed to the symbiogenetic unification of 5-7 different unicellular, ancestral organisms.

An ever-increasing body of data shows that evolution, growth and development – as in all other organismic kingdoms – depends on successful communication processes. This feature is the prerequisite for the internal coordination and organization of the organism and its interplay with other organisms. Communication processes, however, go beyond mere information exchange to include highly differentiated and manifold sign-mediated interactions. The signals used in these interaction processes underly certain semiotic rules. The rule adherence is very reliable and conservative. Nonetheless, the rules governing sign use can be damaged, incompletely executed, deformed, abandoned, or even newly generated. Therefore biosemiotic rules – as opposed to natural laws – are principally changeable.

As in all sign processes (semioses) in living nature, *syntactic* rules determine the relationship of the signs to one another, i.e. provide combination rules for the sequence, combinatory ability, density and rhythm of the signs used. The syntactic rules differ from *pragmatic* rules.

They enable entirely different interactions in that the interaction partners must generate a very specific behavior in order for interactions, for example mutual coordination and organization, to be successful at all. Here, concrete life situations with very specific behavioral contexts are involved. Depending on the context of use, one and the same syntactically correct sign sequence of chemical signal molecules can take on different meanings (*semantic* functions).

Plants fundamentally depend on successful communication. The behavior in the specific interaction can be misinterpreted. A plant can feign mutualism, for example, in order to gain a one-sided advantage from the interaction and to damage, permanently exploit or kill the partner. This, however, cannot be the representative form of communication because no individuals would survive if all plants behaved in this manner. The majority of interactions must be successful for several participants.

Communication processes are successful when the rules governing sign use are correctly followed. Clearly, rules can be broken. In such cases, the messages transmitted via the signs are incomplete, incorrect, and induce no or a false behavioral response. Messages can also be misinterpreted: The sign user uses (a) the sign incorrectly/misleadingly, and the message does not arrive in the manner intended and for the envisioned purpose because it is mutilated, fragmentary. In a due course, the recipient cannot respond the message in the manner required by the non-mutilated message. (b) The sign continuously expresses a message that does not conform with reality ("insect enemies are attacking"); the recipient of the message will respond in a manner adapted to the reality of this inconformity. (c) The message is used to mean something other than it is normally used to convey (in order to gain one-sided advantages). Any constant rule-breaking blocks the organization of life processes (communicative coordination of evolution, reproduction, growth, development) within and between organisms.

The term semiochemicals was generally used to designate molecules or substances (hormones) that served in communication between organisms. As the present review demonstrates all chemicals which function as signs in sign-mediated interactions in and between organisms are semiochemicals. This would be coherent with the biosemiotic approach, which considers the full range of sign use within and between living organisms.[2]

In the future, the meta-, inter- and intraorganismic levels of communication processes will be better understood. This, in turn, will allow better differentiation of the different levels of rules that govern signal use. The syntax of intracellular sign use differs from the syntax at the intercellular level, as well as from the syntax of the signs in species-specific interactions, which in turn differs from sign use in transspecific interactions between organisms. Embedded in the ecological framework, these rules for constituting sign-mediated interactions are used differently depending on the behavioral context. One and the same message can contain, in other contexts, entirely different meanings. Integrating this biosemiotic perspective will help to more gradually decipher the specific meaning of the full range of semiochemicals (in their broader sense) and to get aware of the high level communicative competences of plants.

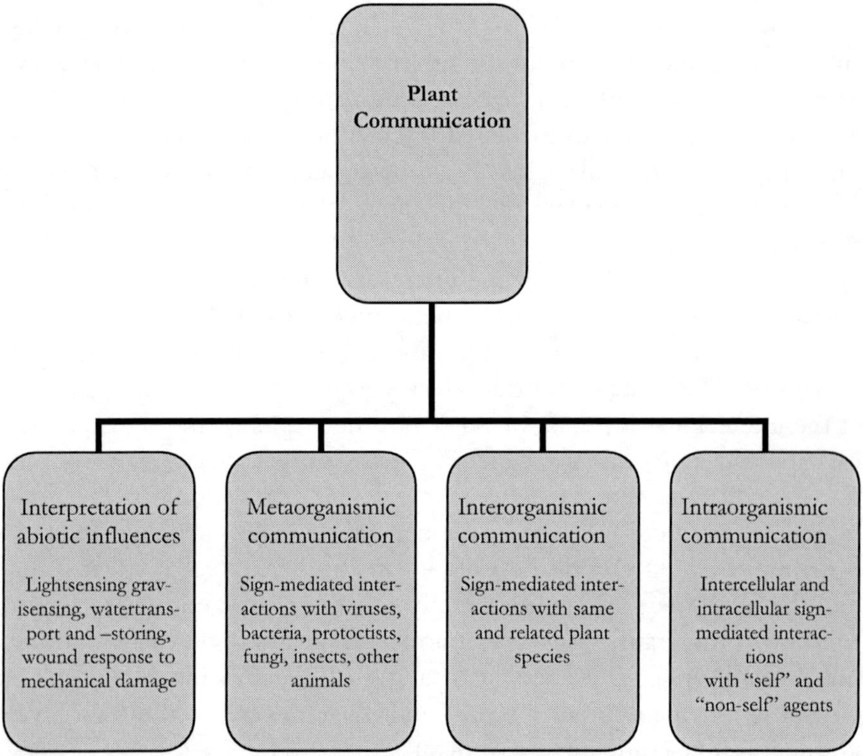

Figure 5: Plant communication processes are sign-mediated interactions which each obey three levels of semiotic rules (syntactic, pragmatic, semantic).

Appendix: Biosemiotics

Biosemiotics (bio = life, Semeion (Greek) = sign)

Biosemiotics is a transdisciplinary science involving theoretical and empirical studies which investigates sign processes (semioses) within and between organisms in manifold communication patterns. Signs may be signals or symbols, most of them chemical molecules, but also physical ones. In highly developed eukaryotic kingdoms, behavioral patterns of organisms may also serve as signs (signals and/or symbols), as for example, the dances of bees; or these signs may be phonetic, as in songbirds or humans.

The signs used obey semiotic rules of three types in principle. Syntactic rules determine combinatory possibilities - physical, chemical, spatial, temporal, rhythmical. Pragmatic rules determine interactional content – e.g., growth, development, defense, mating. And those which are dependent on pragmatic interactional content are the semantic rules, i.e. the meaning function of signs and sign sequences (e.g. in signaling pathways).

Individuals in populations share a common set of signs and a common set of rules. These also apply at the level of cell biology. Dependent on the situational context of interacting entities, signs, or sequences of signs, can have different meanings and functions. Therefore it is possible that different cell types are developed from the same genome trough different chromosomal methylation patterns. Biosemiotics includes not only sign processes used within cells in the context of their molecular and in cell biology, but also embraces immunological, metabolic, neurological and hormonal signalling networks.To many biosemioticians, the origin of life is the starting point of semiosis and vice versa.[1-5]

So far, biosemiotic terms have been used as metaphors in molecular and evolutionary biology, as well as in genetics and ecology, the conviction being that they could ultimately be replaced by chemical and physical descriptions. As a result, the paradigmatic differences between biosemiotic and chemical/physical descriptions are becoming ever more evident and enable biosemiotics to draw a clear distinction between the biotic and abiotic domains: "Life is distinguished from the nonliving world by its dependence on signs".[6] Thereby, it is possible to use biosemiotics to vastly expand the perspective of biological processes. Consequently, the decisive aspects for life processes are not just with respect to the states of matter and their corresponding changes based on natural laws, but the communication and information processes within and among cells, tissues, organs, organisms. Their importance determines the success or failure to promote life, growth, development, disease and death in all living beings.

References

1. Witzany G. 1993. Natur der Sprache – Sprache der Natur. Sprachpragmatische Philosophie der Biologie. Würzburg: Königshausen und Neumann.
2. Witzany G. 2006. The *Logos* of the *Bios* 1. Contributions to the Foundation of a three-leveled Biosemiotics. Helsinki: Umweb.
3. Hoffmeyer J. 1996. Signs of Meaning in the Universe. Bloomington: Indiana University Press.
4. Barbieri, M. 2001. The Organic Codes. The birth of semantic biology. Ancona: PeQuod.
5. Markos A. 2002. Readers of the Book of Life. Oxford, New York: Oxford University Press.
6. Pattee HH. 2005. The Physics and Metaphysics of Biosemiotics. J Biosem 1:281-301.
7. Eigen M, Winkler R. 1975. Das Spiel - Naturgesetze steuern den Zufall. München: Serie Piper.
8. Witzany G. 1995. From the "logic of the molecular syntax" to molecular pragmatism. Evolution and Cognition 1(2):148-168.
9. Bais HP, Park SW, Weir TL, Callaway RM, Vivanco JM. 2003. How plants communicate using the underground information superhighway. Trends Plant Sci 9(1):26-32.
10. Flores HE, Vivanco JM, Loyola-Vargas VM. 1999. 'Radicale' biochemistry: the biology of root-specific metabolism. Trends Plant Sci 4(6):220-226.
11. Baluska F, Mancuso S, Volkmann D, Barlow P. 2004. Root apices as plant command centres: the unique, 'brain-like' status of the root apex transition zone. Biologia (Bratisl) 59(13):7-19.
12. Baluska F, Volkmann D, Menzel D. 2005. Plant synapses: actin-based domains for cell-to-cell communication. Trends Plant Sci 10(3):106-111.
13. Schlicht M, Strnad M, Scanlon MJ, Mancuso S, Hochholdinger F, Palme K, Volkmann D, Menzel D, Baluska F. 2006. Auxin immunolo-

calization implicates vesicular neurotransmitter-like mode of polar auxin transport in root apices. Plant Sign Beh 1:122-33.
14. Campanoni P, Blasius B, Nick P. 2003. Auxin transport synchronizes the pattern of cell division in a tobacco cell line. Plant Physiol 133:1251-60.
15. Casson SA, Lindsey K. 2003. Genes and signalling in root development. New Phytol 158:11-38.
16. Friml J, Wisniewska J. 2005. Auxin as an intercellular signal. In: Fleming AJ. ed. Intercellular Communication in Plants. Annual Plant Reviews. Oxford: Blackwell Publishing 16:1-26.
17. Xia Y. 2005. Peptides as signals. In: Fleming AJ. ed. Intercellular Communication in Plants. Annual Plant Reviews. Oxford: Blackwell Publishing 16:27-47.
18. Mauch-Mani B, Mauch F. 2005. The role of abscisic acid in plant pathogen interactions. Curr Opin Plant Biol 8(4):409-423.
19. Amzallag GN. 2002. Brassinosteroids as metahormones: evidences from specific influence during critical period in Sorghum development. Plant Biol (Stuttg) 4:656-663.
20. Thummel CS, Chory J. 2002. Steroid signaling in plants and insects – common themes different pathways. Genes Dev 16:3113-3129.
21. Fleming AJ. 2005. The plant extracellular matrix and signalling. In: Fleming AJ. ed. Intercellular Communication in Plants. Annual Plant Reviews. Oxford: Blackwell Publishing 16:85-107.
22. Kidner CA, Martienssen RA. 2005. The developmental role of micro RNA in plants. Curr Opin Plant Biol 8:38-44.
23. Farmer E, Schulze-Lefert P. 2005. Biotic interactions: From molecular networks to inter-organismal communities. Curr Opin Plant Biol 8(4):343-345.
24. Torres MA, Dangl JL. 2005. Functions of respiratory burst oxidase in biotic interactions, abiotic stress and development. Curr Opin Plant Biol 8(4):397-403.
25. Delledonne M. 2005. NO news is good news for plants. Curr Opin Plant Biol 8(4):390-396.
26. Apel K, Hirt H. 2004. Reactive oxygen species: metabolism, oxidative stress, and signal transduction. Annu Rev Plant Biol 55:373-99.

27. Carol RJ, Dolan L. 2006. The role of reactive oxygen species in cell growth: lessons from root hairs. J Exp Bot 57:1829-34.
28. Braam J. 2005. In touch: plant responses to mechanical stimuli. New Phytol 165:373-389.
29. Morita MT, Tasaka M. 2004. Gravity sensing and signaling. Curr Opin Plant Biol 7(6):712-718.
30. Hellmeier H, Erhard M, Schulze ED. 1997. Biomass accumulation and water use under arid conditions. In: Bazzaz FA, Grace J. eds. Plant Resource Allocation. London: Academic Press pp. 93-113.
31. Zimmermann u, Schneider H, Wegner LH, Haase A. 2004. Water ascent in tall trees: does evolution of land plants rely on a highly metastable state? New Phytol 162:575-615.
32. Buckley TN. 2005. The control of stomata by water balance. New Phytol 168:275-292.
33. Trewavas A. 2005. Green Plants as Intelligent Organisms. Trends Plant Sci 10(9):413-419.
34. Ballare CL. 1999. Keeping up with the neighbours: phytochrome sensing and other signalling mechanisms. Trends Plant Sci 4:97-102.
35. Baluska F, Volkmann D, Hlavacka A, Mancuso S, Barlow PW. 2006. Neurobiological View of Plants and Their Body Plan. In: Baluska F, Mancuso S, Volkmann D. eds. Communication in Plants. Berlin/Heidelberg: Springer pp. 19-35.
36. Pare PW, Tumlinson JH. 1999. Plant Volatiles as a Defense against Insect Herbivores. Plant Physiol 121:325-331.
37. Kant MR, Ament K, Sabelis MW, Haring MA, Schuurink RC. 2004. Differential Timing of Spider Mite-Induced Direct and Indirect Defenses in Tomato Plants. Plant Physiol 135:483-495.
38. Engelberth, J, Alborn HT, Schmelz EY, Tumlinson JH. 2004. Airborne signals prime plants against insect herbivore attack. Proc Natl Acad Sci USA 101(6):1781-1785.
39. Nürnberger T, Brunner F, Kemmerling B, Piater L. 2004. Innate immunity in plants and animals: striking similarities and obvious differences. Immunol Rev 198:249-266.

40. Peak D, West JD, Messinger SM, Mott KA. 2004. Evidence for complex collective dynamics and emergent-distributed computation in plants. Proc Natl Acad Sci USA 101:981-922.
41. Van der Putten WH, Vet LEM, Harvey JA, Wackers FL. 2001. Linking above- and belowground multitrophic interactions of plants, herbivores, pathogens, and their antagonists. Trends Ecol Evol 16:547-554.
42. Mithöfer A, Wanner G, Boland W. 2005. Effects of Feeding *Spodoptera littoralis* on Lima Bean Leaves. II. Continuous Mechanical Wounding Resembling Insect Feeding Is Sufficient to Elicit Herbivory-Related Volatile Emission. Plant Physiol 137:1160-1168.
43. Pearce G, Ryan CA. 2003. Systemic signaling in tomato plants for defense against herbivores: isolation and characterization of three novel defense-signaling glycopeptide hormones coded in a single precursor gene. J Biol Chem 278(32):30044-30050.
44. Zipfel C, Felix G. 2005. Plants and animals: a different taste for microbes? Curr Opin Plant Biol 8(4):353-360.
45. Nomura K, Melotto M, He SY. 2005. Suppression of host defense in compatible plant-Pseudomonas syringae interactions. Curr Opin Plant Biol 8(4):361-368.
46. Walker TS, Bais HP, Grotewold E, Vivanco JM. 2003. Root exudation and Rhizosphere Biology. Plant Physiol 132:44-51.
47. Romero GQ, Benson WW. 2005. Biotic interactions of mites, plants and leaf domatia. Curr Opin Plant Biol 8(4):436-440.
48. Callaway RM. 2002. The detection of neighbors by plants. Trends Ecol Evol 17:104-105.
49. Dessaux Y. 2004. Biological Communications and Interactions in the Rhizosphere. Presented at "Rhizosphere 2004"-congress, 12.-17. September 2004, Munich/Germany.
50. Dunn AK, Handelsman J. 2002. Toward an understanding of microbial communities through analysis of communication networks. Antonie van Leeuwenhoek 81:565–574.
51. Teplitski M, Robinson JB, Bauer WD. 2000. Plants secrete substances that mimic bacterial N-acyl homoserine lactone signal activities and affect population density-dependent behaviours in associated bacteria. Mol Plant-Microbe Interact 13:637-648.

52. Federle MJ, Bassler BL. 2003. Interspecies communication in bacteria. J Clin Investig 112:1291-1299.
53. Hirsch AM, Bauer WD, Bird DM, Cullimore J, Tyler B, Yoder JI. 2003. Molecular signals and receptors: controlling rhizosphere interacting between plants and other organisms. Ecology 84: 858-868.
54. Sharma A, Ahgal M, Johri BN. 2003. Microbial communication in the rhizosphere: Operation of quorum sensing. Curr Sci 85(8):1164-1172.
55. Jorgensen R. 1993. The origin of land plants: a union of alga and fungus advanced by flavenoids. BioSystems 31:193-207.
56. Zyalalov AA. 2004. Water flows in higher plants: physiology, evolution and system analysis. Russian J Plant Physiol 51:547-555.
57. Jorgensen RA. 2004. Restructuring the Genome in response to Adaptive Challenge: McClintock's Bold Conjecture Revisited. Cold Spring Harbor Symposia on Quantitative Biology 69:349-354.
58. Denison RF, Kiers TE. 2004. Why are most rhizobia beneficial to their plants, rather than parasitic? Microbes Infect 6:1235-1239.
59. Vandenkoornhuyse P, Baldauf SL, Leyval C, Straczek J, Young JP. 2002. Extensive fungal diversity in plant roots. Science 295:2051.
60. Estabrook EM, Yoder JI. 1998. Plant-plant communications: rhizosphere signalling between parasitic angiosperms and their hosts. Plant Physiol 116:1-7.
61. Yoder, JI. 1999. Parasitic plant responses to host plant signals: a model for subterranean plant-plant interactions. Curr Opin Plant Biol 2:65-70.
62. Keyes WJ, O'Malley RC, Kim D, Lynn DG. 2000. Signaling Organogenesis in Parasitic Angiosperms: Xenognosin Generation, Perception, and Response. J Plant Growth Regul 19(2):217-231.
63. Kahmann R, Basse C. 2001. Fungal gene expression during pathogenesis-related development and host plant colonization. Curr Opin Microbiol 4(4):374-380.
64. Imaizumi-Anraku H, Takeda N, Cherpentier M, Perry J, Miwa H, Umehara Y, Kouchi H, Murakami Y, Mulder L, Vickers K, Pike J, Downie A, Wang T, Sato S, Asamizu E, Tabata S, Yoshikawa M, Murooka Y, Wu GJ, Kawaguchi M, Kawasaki S, Parniske M, Hayashi M. 2005. Plastid proteins crucial for symbiotic fungal and bacterial entry into plant roots. Nature 433:527-531.

65. Van West P, Morris BM, Reid B, Appiah AA, Osborne MC, Campbell TA, Shepherd SJ, Gow NAR. 2002. Oomycete plant pathogens use electric fields to target roots. Mol Plant-Microbe Interact 15:790-798.
66. Samaj J, Baluska F, Voigt B, Schlicht M, Volkmann D, Menzel D. 2004. Endocytosis, Actin Cytoskeleton, and Signaling. Plant Physiol 135:1150-1161.
67. Guerts R, Fedorova E, Bisseling T. 2005. Nod factor signaling genes and their function in the early stages of Rhizobium infection. Curr Opin Plant Biol 8(4):346-352.
68. Lammers PJ. 2004. Symbiotic signaling: new functions for familiar proteins. New Phytol 161:324-326.
69. Brundrett MC. 2002. Coevolution of roots and mycorrhizas of land plants. New Phytol 154:275-304.
70. Müller CB, Kraus J. 2005. Symbiosis between grasses and asexual fungal endophytes. Curr Opin Plant Biol 8(4):450-456.
71. Schultz JC, Appel HM. 2004. Cross-Kingdom Cross-Talk: Hormones shared by Plants and Their Insect Herbivores. Ecology 85(1):70-77.
72. Roossinck MJ. 2005. Symbiosis Versus Competition in Plant Virus Evolution. Nat Rev Microbiol 3:917-924.
73. Shackelton LA, Holmes EC. 2004. The evolution of large DNA viruses: combining genomic information of viruses and their hosts. Trends Microbiol 12(10):458-465.
74. Dunoyer P, Voinnet O. 2005. The complex interplay between plant viruses and host RNA-silencing pathways. Curr Opin Plant Biol 8(4): 415-423.
75. Wang MB, Metzlaff M. 2005. RNA silencing and antiviral defense in plants. Curr Opin Plant Biol 8(2):216-222.
76. Dunnoyer P, Voinnet O. 2005. RNA as a signalling molecule. In: Fleming AJ. ed. Intercellular Communication in Plants. Annual Plant Reviews. Oxford: Blackwell Publishing 16: 49-83.
77. Turkington, R, Sackville Hamilton, R, Gliddon C. 1991. Within-population variation in localized and integrated responses of *Trifolium repens* to biotically patchy environments. Oecologia 86:183-192.

78. Tomilov AA, Tomilova NB, Abdallah I, Yoder JI. 2005. Localized hormone fluxes and early haustorium development in the hemiparasitic plant Triphysaria versicolor. Plant Physiol 138:1469-80.
79. Press MC, Phoenix GK. 2005. Impacts of parasitic plants on natural communities. New Phytol 166:737-751.
80. La Cerra P, Bingham R. 2002. The Origin of Minds. New York: Harmony Books.
81. Xoconostle-Cázares B, Xiang Y, Ruiz-Medrano R, Wang HI, Monzer J, Yoo BC, McFarland KC, Franceschi VR, Lucas WJ. 1999. Plant paralog to viral movement protein that potentiates transport of mRNA into the phloem. Science 283:94-98.
82. Baluska F, Volkmann D, Barlow P. 2004. Eukaryotic Cells and their Cell Bodies: Cell Theory Revised. Ann Botany 94:9-32.
83. Gillespie T, Oparka KJ. 2005. Plasmodesmata – gateways for intercellular communication in plants. In: Fleming AJ. ed. Intercellular Communication in Plants. Annual Plant Reviews. Oxford: Blackwell Publishing 16:109-146.
84. Golz JF. 2005. Lessons from the vegetative shoot apex. In: Fleming AJ. ed. Intercellular Communication in Plants. Annual Plant Reviews. Oxford: Blackwell Publishing 16:147-177.
85. Srinivas BP, Hülskamp M. 2005. Lessons from leaf epidermal patterning in plants. In: Fleming AJ. ed. Intercellular Communication in Plants. Annual Plant Reviews. Oxford: Blackwell Publishing 16:225-239.
86. Tassetto M, Maizel A, Osorio J, Joliot A. 2005. Plant and animal homeodomains use convergent mechanisms for intercellular transfer. EMBO Rep 6(9):885-890.
87. Kim I, Zambryski PC. 2005. Cell to-cell communication via plasmodesmata during Arabidopsis embryogenesis. Curr Opin Plant Biol 8(6):593-599.
88. Kurata T, Okada K, Wada T. 2005. Intercellular movement of transcription factors. Curr Opin Plant Biol 8(6):619-624.
89. Bonke M, Tähtiharju S, Helariutta Y. 2005. Lessons from the root apex. In: Fleming AJ. ed. Intercellular Communication in Plants. Annual Plant Reviews. Oxford: Blackwell Publishing 16:199-223.

90. Schilmiller AL, Howe GA. 2005. Systemic signaling in the wound response. Curr Opin Plant Biol 8(4):369-377.
91. McClintock B.1984. The significance of responses of the genome to challenge. Science 226:792-801.
92. Goh CH, Nam HG, Park YS. 2003. Stress memory in plants: a negative regulation of stomatal response and transient induction of rd22 gene to light in abscisic acid-entrained Arabidopsis plants. Plant J 36:240-255
93. Trewavas A. 2003. Aspects of Plant Intelligence. Ann Bot 92:1-20.
94. Coupland G. 2005. Intercellular communication during floral initiation and development. In: Fleming AJ. ed. Intercellular Communication in Plants. Annual Plant Reviews. Oxford: Blackwell Publishing 16:178-197.
95. McCubbin AG. 2005. Lessons in signalling in plant self-incompatibility systems. In: Fleming AJ. ed. Intercellular Communication in Plants. Annual Plant Reviews. Oxford: Blackwell Publishing 16: 240-275.
96. Baluska F, Volkmann D, Barlow PW. 2005. Cell-Cell Channels and Their Implications for Cell Theory. In: Baluska F, Volkmann D, Barlow PW. eds. Cell-Cell Channels. Eurekah.com 1-17.
97. Bell P. 2001. Viral eukaryogenesis: was the ancestor of the nucleus a complex DNA virus? J Mol Ev 53:251-256.
98. Ryan F. 2002. Darwin's blind spot: evolution beyond natural selection. Boston. Houghton Mifflin Company.
99. Villarreal LP. 2005. Viruses and the Evolution of Life. Washington, D.C.: ASM Press.
100. Thompson KR, Olofsdotter OK, Chen DY, Zhao Y, O'Dell TJ, Martin KC. 2004. Signaling from synapse to nucleus during long-term synaptic plasticity: a role for the classical active nuclear import pathway. Neuron 44:997-1009.
101. Martin KC. 2004. Local Protein Synthesis during Axon Guidance and Synaptic Plasticity. Curr Opin Neurobiol 14:305-310.
102. Moccia R, Chen D, Lyles V, Kapuya EEY, Kalachikov S, Spahn CMT, Frank J, Kandel ER, Barad M, Martin KC. 2003. An unbiased cDNA library prepared from isolated Aplysia sensory neuron processes is enriched for cytoskeletal and translational mRNAs. J Neurosci 23:9409-9417.

103. Andersson JO. 2005. Lateral gene transfer in eukaryotes. Cell Mol Life Sci 62:1182-1197.
104. Meagher TR, Vassiliadis C. 2005. Phenotypic impacts of repetitive DNA in flowering plants. New Phytol 168:71-80.
105. Strand A. 2004. Plastid-to-nucleus signalling. Curr Opin Plant Biol 7(6):621-625.
106. Dugas DV, Bartel B. 2004. MicroRNA regulation of gene expression in plants. Curr Opin Plant Biol 7(5):512-520.
107. Bender J. 2004. Chromatin-based silencing mechanisms. Curr Opin Plant Biol 7(5):521-526.
108. Jablonka E, Lamb MJ. 2002. The Changing Concept of Epigenetics. Ann N Y Acad Sci 981:82-96.
109. Rapp RA, Wendel JF. 2005. Epigenetics and plant evolution. New Phytol 168:81-91.
110. Kumar A, Bennetzen JL. 1999. Plant retrotransposons. Annu Rev Genet 33:479-532.
111. Kovalchuk I, Kovalchuk O, Kalck V, Boyko V, Filkovski J, Heinlein M, Hohn B. 2003. Pathogen-induced systemic plant signal triggers DNA rearrangements. Nature 423:760-762.
112. Samaj J, Read ND, Volkmann D, Menzel D, Baluska F. 2005. The endocytic network in plants. Trends Cell Biol 15(8):425-433.
113. Lolle SJ, Victor JL, Young JM, Pruitt RE. 2005. Genome-wide non-mendelian inheritance of extra-genomic information in Arabidopsis. Nature 434:505-509.
114. Weigl D, Jürgens G. 2005. Genetics. Hotheaded Healer. Nature 434:443.
115. Pearson H. 2005. Cress overturns textbook genetics. Nature 434:351-360.
116. Witzany G. 2005. Natural history of life: History of communication logics and dynamics. S.E.E.D. J 5(1):27-55.

2

Fungal Communication*

Abstract: Development and growth of fungal organisms depend on successful communication processes. This review will demonstrate that communication processes within and between fungal organisms occur on three levels: (A) meta-organismic (between fungi and non-fungal organisms); (B) interorganismic (between fungi of the same or related species); (C) intraorganismic (inter- and intracellular). Communication processes are rule-governed sign-mediated interactions. Signs have to be interpreted by the organism to generate in response an appropriate behavior. Signs may be abiotic indicators, biotic signals (chemical molecules or physical contact) or symbols (digital structure of the genome). However, these communication processes can also fail. In such cases the overall results can induce even lethal consequences for the organism.

A biosemiotic perspective investigates communication processes of fungi according to syntactic (combinatorial), pragmatic (context-dependent) and semantic (meaning/function) rules. Thus, Biosemiotics may contribute to a better holistic understanding of these fascinating organisms.

Keywords: Multilevel fungal-communication/sign-mediated interactions/context-dependency

* Gathering in Biosemiotics 6, July 5 – 9, Salzburg 2006

1. Introduction

Most of us enjoy forest mushrooms, i.e., fruiting bodies of fungi, via manifold variation of delicious dishes. However, when fruit, vegetable or bread is coated with a whitish "fur", we know that they are no longer eatable, as some of the poisons produced by these fungi are known carcinogenics. The presence of fungi is even less appreciated when we are confronted by fungal skin infections – they are unpleasant, itchy and require laborious medical procedures. Fungal diseases in the oral cavity, throat or even in internal organs are not just unpleasant, they are dangerous and can be even life threatening although they are closely connected with hominid evolution.[1] In agriculture it is known that fungi are a common cause for recurring disease events on cultivated plants. Also animal farms and maricultural actvities have to deal with fungal diseases.

Although it is known that fungal spores spread via the air, we hardly recognize their omnipresence, mostly because they are hardly seen by the naked eye. Being suspended in the air, they just wait to settle on a suitable substrate, especially when temperature and moisture offer ideal conditions, i.e., to activate the spores, to germinate, to initiate hyphal growth and to develop mycelia.

Referring to evolutionary time scales, the kingdom of fungi emerged approx. 300 million years after the appearance of the first animal species although they descended from a common ancestor.[2,3] On top of that, they share common traits of eukaryotic organisms. In contrast to animal and to higher plant life, monocellular representatives are fairly common among fungi, i.e., fungi are by no means mere multicellular eukaryotes. A fact that can be easily and coherently reconstructed through the lineages of Protoctista is that coordinated behavioral patterns are found among single-celled eukaryotes, which closely resemble those of single-celled fungi. However, there are unmistakable and significant differences in protoctist structure (flagellated) and those of fungi (non-flagellated). Obviously, fungi have evolved out of the coordinating competence of protoctists, such as red and joch algae, respectively.[4]

As plants, fungi are sessile organisms that can live for extremely long periods or extend over large areas: one exemplar has been found which covers as much as 15 hectares with an age structure of approximately 1500 years. Endolithic fungi of the Antarctic are known to be among the most long-lived organism on this planet.[5]

Most fungi are saprobes and decompose and feed on non-living organic matter. They secrete powerful enzymes that enable the cells to digest organic matter outside of their body in the nearby environment, which in turn is broken down to smaller molecules that can be absorbed and reincorporated in dissolved form. To deter potential predators, a number of complex and highly efficient deterring substances are produced by fungi.[4]

For 1000s of years, fungal competences are actively utilized by humans in beer, wine, and bread production. Since fungi are very simple organisms, sequencing and technical manipulation are relatively easy, making them ideal organisms for laboratory experiments[6] such as *Neurospora grassa*[7,8] or yeasts. For example, recent research on yeast genomes has demonstrated important features of higher order regulatory functions hidden in their DNA. Studies have showed that DNA encodes its own packaging, i.e., DNA sequences code how to package their own genome in the nucleosomes.[9]

Although fungal infections cause a wide range of diseases, they serve also as important resource for drug production. Their application in the medical field as producers of various antibiotics is comprehensive and well appreciated.[10]

2. Semiochemical vocabulary

Since coordination and organization processes occur in all organismic kingdoms, fungi are no exception: there are such processes during the formation of mycelia and fruiting bodies (intraorganismic), between species of same kind (interorganismic) and between fungi and non-fungal organisms (metaorganismic). The processes are subjected to rule-governed sign-mediated interactions. These signalling processes are nothing else but distinct biosemiotic communication processes.[11] The used semiochemicals[12] are of biotic origin and they trigger the fungal organism to react in a specific manner. It means, fungal organisms are competent of identifying the difference of the same molecules being part of biotic messages or not.

So far five different primary signalling molecules are known that serve to coordinate very different behavioral patterns such as filamentation, mating, growth, pathogenicity.[13-17] Behavioral coordination and the production of such substances can only be achieved through interpretation processes: self or non-self, abiotic indicator, biotic message from similar, related, or non-related species, or even "noise", i.e., similar molecules without interpretational content. Furthermore, there are numerous, lesser-investigated subunits that attain an accompanying role as they are weaker in effect.[2] Globally, these semiochemicals serve to coordinate similar goals in different fungal species, yet species-variation among them cannot be ignored.

These patterns regard:[18]

- mitogen activated protein kinase signalling (MAPK) serves in diverse processes: cell integrity, cell wall construction, pheromene/mating and osmo-regulation;

- the cyclic adenosine monophosphate cAMP/PKA system serves in fungal development and virulence;
- the RAS protein serves for cross-talking between signalling cascades;
- calcium-calmodulin-calcineurin serves for cell survival under oxidative stress, high temperature, membrane/cell wall perturbation;
- rapamycin serves for control of cell growth and proliferation.

Up to now 400 different secondary metabolites have been documented. These are known to contain mycotoxins and are used both for defensive and aggressive behaviorial patterns.

3. Internal interpretation of external abiotic influences

Fungi react sensitively to varying nutrient availability and nutrient fluxes, responding to initial intraorganismic communication. That is, in case of carbon or nitrogen insufficiencies, the internal communication of the organism responds adequately and is phenotypically expressed by a change in hyphael growth. Up to now, two specific signalling pathways have been found that coordinate such behavior. These diverging patterns have also been documented among other fungi, including those that are pathogenic to plant and animal life.[2]

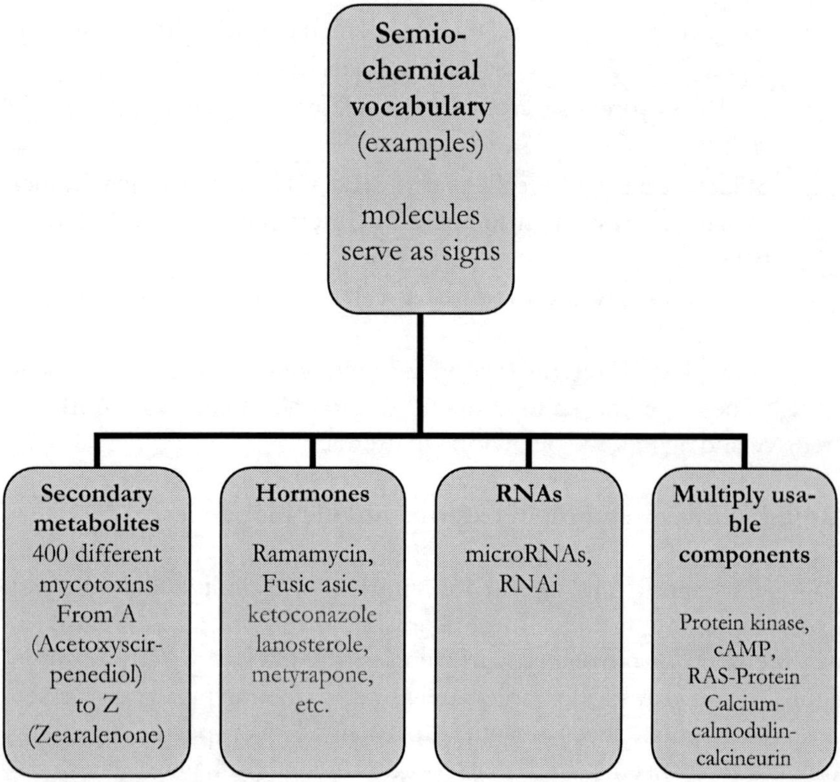

Figure 1: Examples of semiochemicals in fungal communication processes

Abundance of nutrients for example, results in an increased production of cAMP, which suppresses that of Ste11, itself a mating inhibitor. Such processes are sign-mediated processes that originate from cyclase associated proteins (CAP).[2]

As animals and plants, seasonality as a part of the circadian system is also found in fungi.[8] In particular, it regards light regulated physiologic processes that coordinate the internal fungal clock.[19] *Neurospora crassa* for example, responds according to irradiance patterns as a result of the diurnal cycle. In such cases we can speak of a Neurospora chronotype.[8,20] However, these responsiveness is not connected to photosynthetic activity, as this is only found in green algae, i.e., the symbiotic association of lichens.

Most fungi are saprobes and feed on dead organic matter. Almost all fungi digest food outside their bodies: the excretion of extracellular digestive enzymes fragments larger biomolecules and makes them soluble, which are then readily accessible for the fungal organism. This is particularly important for the digestion of cellulose, through enzymatic activity of exocellulase and endocellulase, lignin (lignin peroxidase and manganese peroxidase). Enzymatic breakdown of organic matter yields simple sugars, amino acids, fatty acids and other smaller molecular components.[4]

4. Metaorganismic (trans-specific) communication

Fairly early associates of the fungal kingdom are those organisms that interact symbiotically with fungi, as is the case with lichens. All higher fungal life-forms originated from these symbiotic ways of life, which later became independent by detaching themselves from this close and vital dependence. Lichens are complementary symbiotic partnerships between photobionts (algae or bacteria with fungi), i.e., they are viable only together.[21] As pioneering organisms, they may settle on blank rock. Being essential bioeroders, they extract nutrients from mineral matter, thereby initiating the process of soil formation and paving the way for successive organisms that include even root-forming species.

Fungi are known to utilize a broad variety of different symbiotic interactions with animals, plants and eukaryotic protozoa for both mutual benefit as well as parasitic and even lethal associations. They also settle on specific types of tissue. Fungal diseases are known to affect both plant and animal life where they can induce devastating effects in agriculture.

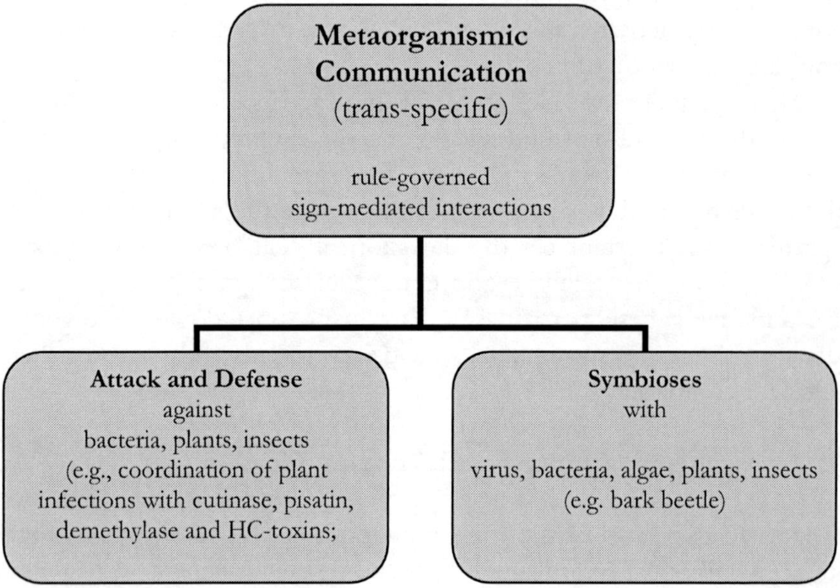

Figure 2: Metaorganismic (trans-specific) fungal interaction patterns

A typical example of mutually beneficial symbiosis can be found between the bark beetles and quite a few different fungi.[12] The fungal spores benefit from the locomotion provided by the beetles in several aspects: access to new hosts, while the beetle benefits by the availability of fungal nutrients and pheromones. Some fungi provide nitrogen, amino acids and sterols, that are crucial for the development of beetle larvae, however, this gives effect only once the adult beetle has colonized a host. Many bark beetles even evolved transportation pockets for fungal hyphae, that points at the common evolutionary history.[22]

Another metaorganismic symbiotic signalling happens between fungi and ants which derives from a co-evolutionary relationship which lasted for over millions of years.[23] Interestingly, some Lignin-degrading fungi also produce semiochemicals which have effects on the feeding and foreaging behavior of a Formosan subterranean termite.[24]

Many of the known indicators stress the fact that both the fungal as well as the animal kingdoms share common ancestors, such as protoctists with a true nucleus like choanoflagellates.[5] With reference to this, fungi and animals are more related to each other than to the plant kingdom. This is further strengthened by the sign-mediated processes, which regulate cellular functions. Yet, a different indicator of their common ancestor is found in a particular signalling pathway, termed mitogen activated protein kinase cascade (MAPK). It plays a crucial role in cell wall stabilization of fungi and pheromone/mating interaction among mammal cells. On top of that, MAPK is highly conservative.[2]

Then, there are fungi, which parasitize plants. For example, they colonize host tissues with an intercellular mycelium that forms haustoria, i.e., fungal mats within plant cells[25] that penetrate the cell to utilize the nutrients of the plant. Investigations of hazardous fungal infections on plants revealed the crucial role of enzymes such as cutinase, pisatin, demethylase and HC-toxins.[26]

Today, several hundred species of fungi colonize more than 100,000 different plant species. This type of cohabitation requires symbiotic signalling[27], e.g. the starting of filamentous growth of fungi through plant hormones.[28]

Roots of plants provide better conditions for mycorrhizal fungi, which in turn supply plants with better nutrients.[29] For the fungus, such a relationship is either balanced or predatory. Endophytic fungi, however, live in plants without triggering symptoms of disease.[29] Today, scientists consider the origin of plant cell to be the result of terrestric activity of mycorrhiza; i.e., settlement on land is a coevolutive event that is comparable in the mode of complementarity to that between flowering plants and insects.[5] Thereby fungi excrete digestive enzymes into the surrounding soils, convert nutrients into aqueous solutions that in turn can be readily absorbed by the plant.[30] Therefore, a staggering majorities of 80% of all terrestrial plants, especially trees, rely on the activity of mycorrhiza.[31]

Fungi affecting animals are usually dependent on the host's body temperature; i.e., only if the body temperature is sufficiently high, host colonization by the fungi is possible. Especially *Aspergillus Fumigatus* colonizes animal hosts if they are under thermal stress.[32] Although fungal disease is common in birds, the relative resistance of endothermic vertebrates to fungal diseases may be a result of immune responses connected with higher body temperature.[33]

A typical fungi cultivated in research laboratories is Neurospora. Several subspecies of this genus contain toxins that are pathogenic for animals and plants. However, they are essential components for large-scale industrial production of antibiotics, chemicals, enzymes and pharmaceuticals.[8]

One of the most striking metaorganismic communication processes between fungi and non-fungal species can be found in lichens. Lichens are polyphyletic. They have been derived many times independently from different kinds of ascomycetes, so undoubtedly, their nature of symbiosis varies.[34] In lichens the algae provides the fungi with photosynthate, while the fungi caters the algae with nutrients. Lichens constitute one of the oldest known fungal members and are capable to resist quite averse environmental conditions.

The symbiosis between fungi and algae or fungi and bacteria result in a mutual supply of nutrients and their associated competences. Through quorum sensing, the fungi benefit from the bacterial association.[35] In turn, the bacteria utilize dissolved fungal metabolites to satisfy their nutritional requirements. A similar co-dependence is observed with algae as symbiosis partners.

Metaorganismic communication can be found in any interaction that involves fungi and viruses, bacteria, protoctista (algae), animals (insects) and plants.

5. Interorganismic (species-specific) communication

Since there are both single and multicellular species, determination of communication processes between same species and related fungal species cannot be distinguished unambiguously from intercellular communication (intraorganismic). Thereby communication processes of monocellular yeast,[36] which resembles that of amoeba like Dictyostelium, must be considered as interorganismic communication. Whereas, cell-to-cell communication between multicellular fungal species is truly intercellular. To verify whether cell-to-cell communication is inter- or intraorganismic, we have to consider intercellular processes on a case-by-case basis.

Herein lies another fundamental characteristic of biota: it regards the identification competence for the determination of "self" and "non-self". This competence could be successfully established in *Neurspora crassa.*[37-39] It is obvious that this capacity to distinguish between oneself and the others is vital for fungi; that is, the encounter of mycelia among same species results in the merger of their fungal hyphae. However, such dikaryotic mycelia can also result from the merger of different fungal species. While peripheral hyphae tend to avoid merging with hyphae of other species, the opposite is the case with those at the center of the mycelium.[37-39]

If one assigns mycelia the role of a wrapper within which the fungi, so to speak, is enveloped into a fluid-like continuum, nuclei of compatible but different species are then “flowing” through the same mycelium. The overall result is an organism, which houses nuclei of different genetic origin in its cytoplasm.[40] However, if specific genetic sequences are incompatible, then repulsion sets in forcing the approaching hyphae to an immune-like response.

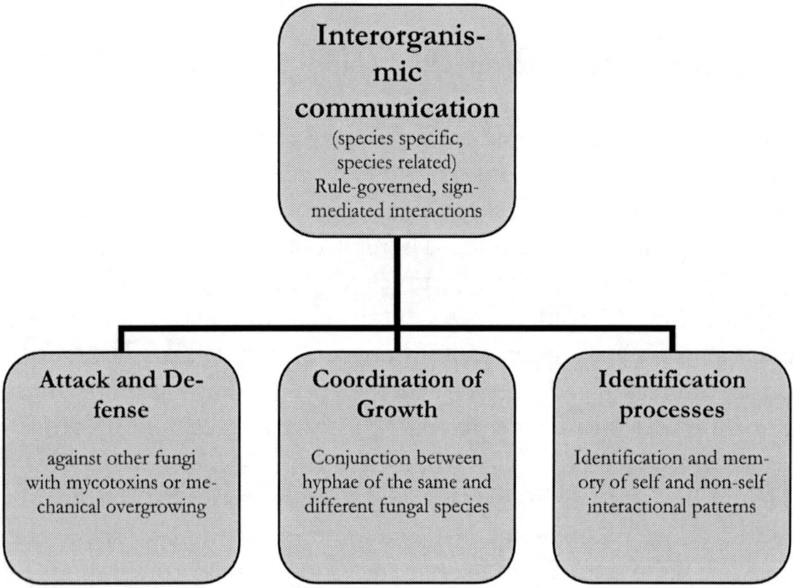

Figure 3: Examples for behavioral contexts of interorganismic fungal communication

Resource competition in fungi occurs directly, indirectly and via mechanic interaction. Indirect competition involves absorption of all available resources within the reach of the mycelium, thereby famishing potential competitors by maintaining a nutritional deficiency gradient. Direct interaction, on the other hand, involves secondary metabolites, which suppress growth or even induce death of the competing fungi. The antibiotics employed in such cases can be either volatile or non-volatile. Mechanical interaction simply requires overgrowth of one fungal species by the other, in which the overgrowing species exerts its lytic action on the other. In some cases lysis is induced via antibiotical agents.[41,42]

As bacteria are also single-celled, fungi use quorum sensing to regulate and affect biofilm formation and pathogenesis.[43] This is mediated by small molecules which accumulate in the extracellular environment. If it reaches a sufficiently high concentration a response regulator is activated within the local population of cells leading to the coordination of a special gene expression.[35]

In parasitic interactions between fungal organisms even cytoplasmic fusion during infection processes was found which indicated genetic transfer in the host parasite relationship. The recognition pattern in this predatory-prey relationship is mediated by trisporoids which also serve in a non-parasitic behavioral pattern, i.e., is responsible for sexual communication.[44]

6. Intraorganismic communication

The countless variety of fungal organisms represents a major challenge when establishing a homogeneous designation of the sign processes employed. Research activities so far have predominantly focused on those fungal species that posed a serious threat to agriculture, are pathogenic to humans or possessed antibacterial properties. Species of this kind are relatively well investigated, whereas species with obviously insignificant properties are hardly known. Growth in *Saccharomyces cerevisiae*, for example is entirely dependent on the final purpose; i.e. its growth and behavioral patterns are correlated: mating of haploid fungal cells is a completely different behavioral cycle than filamentous growth of diploid cells. The signs involved can be differentiated by four different sign-related production patterns.[2,45] There is a new evidence that also proteins which have been restricted to signalling pathways in animals so far are common in fungal signalling.[46]

6.1 Intercellular communication

Higher Fungi are modular hyphal organisms in that they reproduce by clonation or also parasexually. They establish interlocked networks. Like red algae they "merge" their cytoplasms to form multi-nucleated cells. This holistically "merged" body is also found among some plants.

A spore germinates under appropriate environmental conditions and is followed by the formation of filaments called hyphae. The latter is characterized by nuclear division and spore formation, which develops into monokaryotic filaments (tip growing). There is a need, as in other tip growing organisms, in high density of tip-localized mitochondria.[47]

The embryological stage, a characteristic of higher plants and animals, is completely absent in fungi.[4] Hyphae formation is also found among certain bacteria, like Streptomyces and Actinomyces. Hyphae tightly packed together into a mat is called a mycelium. Each filament of hyphae has tubular side walls made mostly of chitin, a feature that is common also in Arthropoda. The cell walls that seem to separate adjacent cells in a filament are called septa; however, their porous nature does not really assign them separating properties. The merger of filamental tips of the same or different species triggers a self / non-self identification process. This process is sign-mediated and results either in repulsion or attraction. If the latter occurs, merger to a dikaryotic mycelium takes place and initiates the formation of a fruiting body.

Hyphal growth is a totally different pattern of conduct than normal cell growth: such cells change shape, become long, and re-orientate themselves into specific directions to come into physical contact or even "merge" with each other only to colonize a potential growth resource – usually dead organic matter.

Fungal hyphae simultaneously extend into a given direction only when nutritional resources are ideally distributed. However, this is a rare event. Usually the fungi propagate in the conditions rich with organic matter (carbon and its derivatives) just to halt growth when little or no resources are available; here the fungi coordinate their growth by employing certain properties. In order to do so, the fungi use intercellular signals that enable them to comprehend the overall state of the organism. This includes an integrated ability to discriminate between self and non-self. Once the fungus encounters a resource-depleted substrate or even poisonous compounds, it responds by halting its growth cycle or by propagating in another direction.[48] The protein signals involved in such processes are quite complex: the apex of hyphae houses specialized receptors that are able to respond to any environmental condition. Any carbon-enriched substrate makes these receptors active, which in turn results in the production and release of protein signals into the hyphal cytosol where the corresponding signalling cascade is triggered. In turn, the mycelium responds with the mobilization and translocation of resources into the activated area. In the absence of carbon-rich substrates, or at increasingly acidic pH levels, the hyphae responds by activating yet other receptors that slow down growth and eventually make the organism withdraw resources from the affected area. Since septa within the hyphae are perforated, they perform similar functions as observed with the gap junctions in higher animals, micro-plasmodesmata in cyanobacteria or plasmodesmata in higher plants.[48]

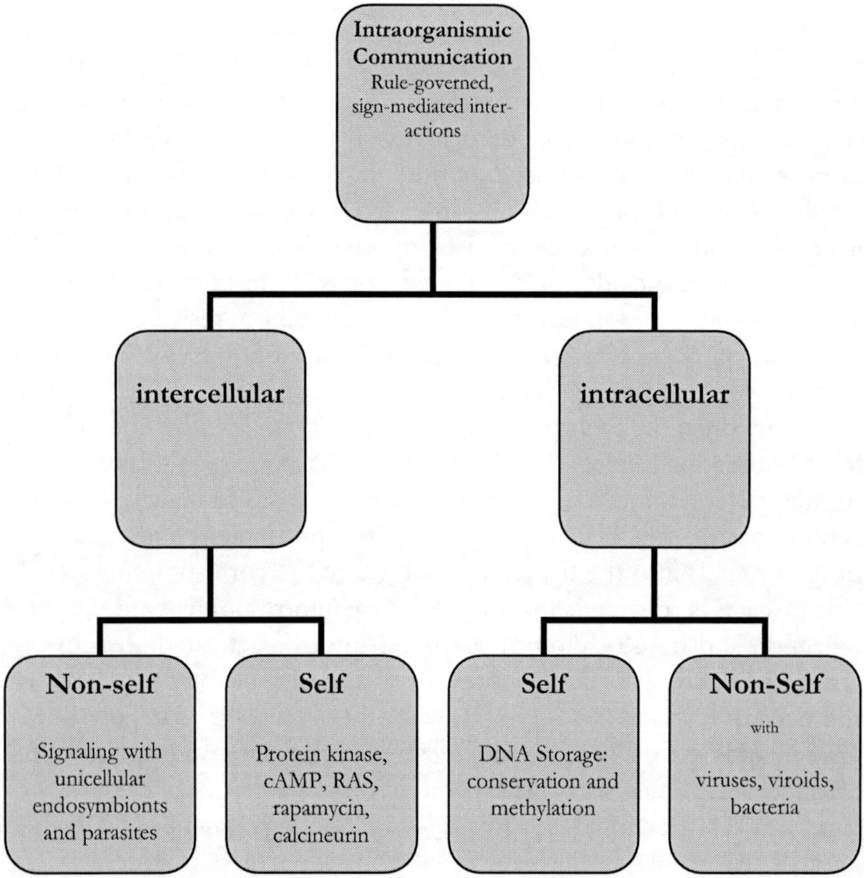

Figure 4: Levels of intraorganismic fungal communication

6.2 Intracellular communication

Unfortunately, our understanding of fungal communication processes for the assembly and merger of cell walls, of capsules, or cellular surface-receptors is incomplete. The secreted signalling molecules that unhinderedly penetrate the cell walls are hardly known. Even the phenomenon of endocytotic absorption of dissolved organic matter into the fungal cell is not really clear.

Yet, there is some progress in this field. By investigating a great number of signal transduction events from the outside through the cell membrane into the cytoplasm, it was possible to decode some important intracellular communication processes. Thereby, it has been found that sign processes coordinate cell polarity, mating, pheromone control and cellular morphology. Some of these sign processes even adjust the cell cycle, perform polarized growth activity and modify the transcription profiles of fungal cells.[2,49,18] By examining the fungal pathogen *Paracoccidioides brasilienus* it has been revealed that some signalling pathways are identical to those of other species such as in *Saccharomyces cerevisiae, Cryptococcus neoformans, Candida albicans* and *Aspergillus fumigatus.*[18]

The involved "protein cascades" that characterize production pathways of appropriate chemicals and messenger signals regard behavioral contexts, which are to some extent completely different:[18]

- cell integrity, cell wall construction, pheromene/mating, Osmoregulation by mitogen activated protein kinase signalling (MAPK);
- fungal development and virulence by the cAMP/PKA system;
- cross talking between cascades by the RAS protein;
- cell survival under oxidative stress, high temperature, membrane/cell wall perturbation by calcium-calmodulin-calcineurin;
- control of cell growth and proliferation by rapamycin.

In addition, combinatorial communication procedures such as MAPK and cAMP pathways are also part of the behavioral contexts.[2] This in turn serves to multiply the semantic contents of the encoded messages.

To sense extracellular stimuli and convert them into intracellular signals which regulate developmental and growth processes several signalling pathways have been found. The guanine nucleotide-binding protein (G-Protein) is essential for extracellular detection of nutrients and sexual partners.[50-52] The TOR protein kinases which are bind and inhibited by rapamycin function as nutrient-sensing signals and regulate cellular responses like proliferation, transcription, translation, autophagy and ribosome biogenesis.[53-56]

As with any sign-mediated interaction that can be achieved with molecules, the same components are employed for different behavioral contexts and in various messages. That is, different modes of behavior can be coordinated by syntactic identical molecules. The signalling pathways use identical proteins to coordinate different response patterns. Even if they are syntactically identical they have completely different meaning. For example, activated cAMP triggers filamentous growth in *Sacharomyces cerevisiae*, regulates positive virulence in *Cryptococcus neoformans*, suppresses mating in *Schizosacharomyces pombe* and inhibits filamentous growth in *Ustilago maydis*[2, 57] or activates protein kinase for directly or indirectly indicated developmental changes in *Magnaporthe grisea* during infection of rice.[58] Another example is the *Ustilago maydis* pheromene response which regulates both the cell fusion as well as the pathegenicity program for plant infection.[59-61]

These examples show that different behavioral contexts determine different meanings of identical signalling molecules, or in biosemiotic terms, *pragmatics determines semantics of syntactic identical substances.* In such cases, biosemiotically identical signs induce opposite responses in their life-related relations in different organisms. It is interesting to note, how fungi are not just capable to differentiate various messages and to respond appropriately, but moreover, to differentiate molecules that are chemically identical from molecules which obviously contain no relevant meaning („noise“), i.e., they are not parts of biotic messages.

Recent genome comparison has given new insights in evolutionary aspects of fungi. The thesis that evolution happened through whole genome duplication events, followed by selective gene-loss and stabilization, is strengthened by analysis on *Saccharomyces cerevisiae*.[62] Interestingly, the signal-to-noise ratio in yeasts is approximately 70% protein coding regions and 15% regulatory elements in the non-protein coding regions in comparison to humans with 3% and 97%, respectively.[63]

Horizontal gene transfer is a common trait in bacterial interactions to share a wide range of genetic capabilities. As recent research suggests this phenomenon is also part of a interorganismic-intracellular communication process between fungi of different species known as plant pathogens:[64,65] Oomycetes and similar fungal organisms produce similar plant cell-wall degrading proteins.[66]

6.3 Unique relationship between fungi and viruses

Fungal populations are commonly colonized by double-stranded RNA viruses, single-stranded RNA viruses, double-stranded DNA viruses, retroviruses and even prions. They are the simplest organisms which represent colonization-events by retroviruses, documented in fungal retroposons. In contrast to all other eukaryotes, fungal mitochondria are infected by either double-stranded RNA viruses or DNA viruses. Most fungus infecting viruses are not pathogenic although they are connected with toxin genes which are pathologically dangerous not for the infected host but for non-infected fungal relatives. This phenomenon has been used quite successfully by brewing industries: to protect useful yeasts from exogenous yeast genetic parasites the industrial yeast colonies have been colonized with protective versions of killer viruses.[5]

An example for mutually beneficial symbiosis is that of black aspergillus species and dsRNA viruses as in the case of mycovirus.[67]

The interconnected filamentous fungi are a very attractive habitat for viruses because they allow rapid motility through the whole fusioned cellular network. Fungal fusioned networks allow also replacement of nuclei which can leave mitochondria behind. This may be an advantage if mitochondria are colonized by parasites as is common the case.[5] It has been suggested that non-pathogen viral colonization of fungal mitochondria protects mtDNA from age-dependent degradation and is beneficial for the longevity of the fungal host. The consequence is that the growth radius of the fungal mats may raise up to 10-fold.[5]

Interestingly, higher fungi lack repetitive elements in DNA, lower fungi can have up to 50% of their genomes colonized by repetitive elements. Recent research shows that these repetitive elements represent remnants of early viral genomic colonization events. In fungi we can find the first colonization by viral RNA. This assumption is underlined by the presence of SINEs and LINEs. So far up to 15 SINE-families have been recognized;[68] a fact that supports the assumption that a very early viral colonization of the fungal genome must have taken place.[5,69]

Although plants and animals have the same descendants as fungi, the relationship between fungi and viruses is different to virus-animal or virus-plant relationships.[5] The latter ones lack linear plasmids, killer phenotypes, mitochondrial infections, distorted senescence and ubiquitous double-stranded RNA colonization which are all characteristics for the fungal virus-host relationship. This may be an indication for coevolutive interactions between fungi and viruses.[5]

Genomic studies show that the most methylated regions of *Neurospora crassa* are derived from repetitive and induced point mutations (RIP) which is a pre-meiotic homology-based genome defense system. This early immune system consists of a variety of inactivated transposons which include DNA modifications and chromatin modifications.[70,71] As we know through comparative genomics, such inactivated transposons descend from viral infection events which have reached a persistent status providing the host with a new phenotype of an innate immune system to prevent similar infections.[5]

7. Discussion

An overview about significant levels of fungal communication shows that identification of sign-mediated processes in signalling pathways are context dependent - both within and among fungal cells as well as between fungi and other organisms.

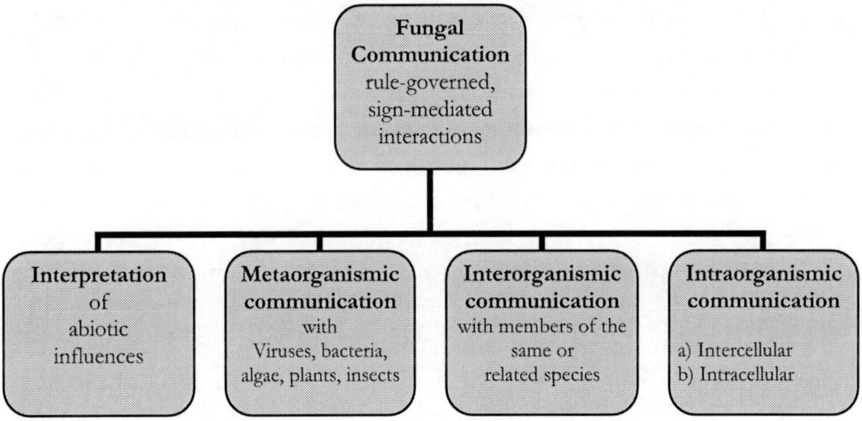

Figure 5: General levels of fungal communication

Such dependence is prevalent in either (beneficial or parasitic) colonization or defense responses. Depending on the utilization context, molecular components are integrated into unique signalling pathways where they gain the corresponding meanings. Such meanings are subject to change; i.e., they rely on various behavioral contexts, which differ in altering conditions. These contexts concern cell adhesion, pheromone response, calcium/calmodulin, cell integrity, osmotic growth, stress response or cell growth through rapamycin. The interactional context (pragmatics) determines the semantic relation, i.e., its meaning and the function of the chemical components, and is found to be a sign-mediated communicational pattern of fungi. This is a common feature in all eukaryotic kingdoms: the context determines the meaning of meta-, inter- and intraorganismic (inter- and intracellular) communication, while differences in abiotic and biotic signal perception determine the content arrangement of responsive behavior.[72]

After recognizing how versatile fungal communication competences really are, we can see that a main principle is followed throughout all these signalling processes: with a core set of chemical molecules fungal organisms coordinate all their behavioral patterns. The interactional (pragmatic) context and the different modes how to coordinate appropriate response behavior, e.g. in development, growth, mating, attack, defense, virulence, etc., determine combinations of signals to generate the appropriate meaning-function, i.e., informational content of messages. In contrast to former systematizations which have been investigating combinatorial rules of meaning and functions in signalling molecules, the biosemiotic perspective differentiates all three levels of rules involved in signalling, the syntactic level (combinations), the pragmatic level (context) and the semantic level (content). These generating processes normally function in a very conservative way but under certain circumstances may fail, or selective pressure may lead to changes which can be a driving force in fungal evolution.

References

1. Lott TJ, Fundyga RE, Kuykendall RJ, Arnold J. 2006. The human commensal yeast, Candida albicans, has an ancient origin. Fungal Genet Biol 42: 444-451.
2. Lengeler KB, Davidson RC, D'Souza C, Harashima T, Shen WC, Wang P, Pan X, Waugh M, Heitman J. 2000. Signal Transduction cascades regulating Fungal Development and Virulence. Microbiol Mol Biol Rev 64: 746-785.
3. Lang BF, O'Kelly C, Nerad T, Gray MW, Burger G. 2002. The closest unicellular relatives of animals. Curr Biol 12: 1773-1778.
4. Margulis L, Schwartz KV. 1988. Five Kingdoms. W. H. Freeman and Company, New York.
5. Villarreal LP. 2005. Viruses and the Evolution of Life. ASM Press, Washington, D.C.
6. Weld RJ, Plummer KM, Carpenter MA, Ridgway HJ. 2006. Approaches to functional genomics in filamentous fungi. Cell Res 16: 31-44.
7. FGSC- Fungal Genetics Stock Center, 2005. The Neurospora Homepage. http://www.fgsc.net/Neurospora/neurospora.html.
8. Dunlap JC, Loros JJ, Denault D, Lee K, Froelich A, Colot H, Shi M, Pregueiro A. 2004. Genetics and molecular biology of circadian rhythms. In The Mycota III. Biochemistry and Molecular Biology, Second Edition, Brambl R, Marzluf GA (eds) pp 209-22, Berlin-Heidelberg: Springer.
9. Ercan S, Lieb JD. 2006. New evidence that DNA encodes its packaging. Nat Gen 38: 1104-1105.
10. Arora DK. 2003. Handbook of Fungal biotechnology. Marcel Dekker, New York.
11. Witzany G. 2006a. The Logos of the Bios 1. Contributions to the Foundation of a three-leveled Biosemiotics. Umweb, Helsinki, Finland.
12. Sullivan BT, Berisford CW. 2004. Semiochemicals From Fungal Associates of Bark Beetles May Mediate Host Location Behavior of Parasitoids. J Chem Ecol 30: 703-717.

13. Adachi K, Hamer JE. 1998. Divergent cAMP signalling pathways regulate growth and pathogenesis in the rice blast fungus Magnaporthe grisea. Plant Cell 10: 1361-1374.
14. Wang P, Heitman J. 1999. Signal transduction cascades regulating mating, filamentation and virulence in Cryptococcus neoformans. Curr Opin Microbiol 2: 358-362.
15. Hemenway CS, Heitman J. 1999. Calcineurin: structure, function, and inhibition. Cell Biochem Biophysics 30: 115-151.
16. Alspaugh JA, Cavallo LM, Perfect JR, Heitman J. 2000. Ras1 regulates filamentation, mating and growth at high temperature of Cryptococcus neoformans. Mol Microbiol 36: 352-365.
17. Borges-Walmsley MI, Walmsley AR. 2000. cAMP signalling in pathogenic fungi: control of dimorphic switching and pathogenicity. Trends Microbiol 8: 133-141.
18. Fernandes L, Araujo MAM, Amaral A, Reis VCB, Martins NF, Felipe MS. 2005. Cell signalling pathways in Paracoccidioides brasiliensis – inferred from comparism with other fungi. Genet Mol Res 4: 216-231.
19. Bell-Pederson D, Dunlap JC, Loros JJ. 1996. Distinct cis-acting elements mediate clock, light, and developmental regulation of the Neurospora crassa eas (ccg2)gene. Mol Cell Biol 16: 513-521.
20. Tan Y, Merrow M, Roenneberg T. 2004. Photoperiodism in Neurospora Crassa. J Biol Rhythms 19: 135-143.
21. Sanders WB. 2001. Lichens: the interface between mycology and plant morphology. BioScience 51: 1025-1035.
22. Kopper BJ, Klepzig KD, Raffa KF. 2004. Components of Antagonism and Mutualism in Ips pini – Fungal interactions: relationship to a Life History of colonizing Highly Stressed and Dead Trees. Environ Entomol 33: 28-34.
23. Poulsen M, Boomsma JJ. 2005. Mutualistic Fungi Control Crop Diversity in Fungus-Growing Ants. Science 307: 741-744.
24. Cornelius ML, Bland JM, Daigle DJ, Williams KS, Lovisa MP, Connick WJ, Lax AR. 2004. Effect of a Lignin-Degrading Fungus on Feeding Preferences of Formosan Subterranean Termite (Isoptera: Rhinotermitidae) for Different Commercial Lumber. J Econ Entomol 97: 1025-1035.

25. Jakupovic M, Heintz M, Reichmann P, Mendgen K, Hahn M. 2006. Microarray analysis of expressed sequence tags from haustoria of the rust fungus Uromyces fabae. Fung Gen Biol 43: 8-19.
26. NIAID (National Institute of Allergy and Infectious Diseases) 1993. Molecular Medical Mycology SECRETION. Proceedings of the NIAID Workshop in Medical Mycology. University of Minnesota, Minneapolis, http://www.niaid.nih.gov/dmid/meetings/mycology/secretion.htm
27. Lammers PJ. 2004. Symbiotic signalling: new functions for familiar proteins. New Phytol 161: 324-326.
28. Prusty R, Grisafi P, Fink GR. 2004. The plant hormone indoleacetic acid induces invasive growth in Saccharomyces cerevisiae. Proc Natl Acad Sci USA 101: 4153-4157.
29. Brundrett MC. 2002. Coevolution of roots and mycorrhizas of land plants. New Phytol 154: 275-304.
30. Walker TS, Bais HP, Grotewold E, Vivanco JM. 2003. Root exudation and rhizosphere biology. Plant Physiol 132: 44-51.
31. Schwarze FWMR, Engels J, Mattheck C. 2004. Fungal Strategies of Wood Decay in Trees. Second Edition, Heidelberg: Springer.
32. Bhabhra R, Zhao W, Rhodes JC, Askew DS. 2006. Nucleolar localization of aspergillus fumigatus CgrA is temperature-dependent. Fungal Genet Biol 43: 1-64.
33. Casadevall A. 2006. Fungal virulence, vertebrate endothermy, and dinosaur extinction: is there a connection? Fungal Genet Biol 42: 98-106.
34. Raven JA. 2001. Selection pressures on stomatal evolution. New Phytol 153: 371-386.
35. Hogan DA. 2006. Talking to Themselves: Autoregulation and Quorum Sensing in Fungi. Eukaryot Cell 584: 613-619.
36. Banuett F. 1998. Signalling in the yeasts: an informational cascade with links to the filamentous fungi. Microbiol Mol Biol Rev 62: 249-274.
37. Glass NL, Jacobson DJ, Shiu PKT. 2000. The Genetics of Hyphal Fusion and Vegetative Incompatibility in filamentous ascomycete Fungi. Annu Rev Genet 34: 165-186.
38. Glass NL, Saupe SJ. 2002. Vegetative incompatibility in filamentous ascomycetes. In Molecular Biology of Fungal Development Osiewacz HD (ed) pp 109-131 New York: Marcel Dekker.

39. Glass NL, Kaneko I. 2003. Fatal Attraction: Nonself Recognition and Heterokaryon Incompatibility in Filamentous Fungi. Eukaryot Cell 2: 1-8.
40. Wu J, Glass NL. 2001. Identification of specificity determinants and generation of alleles with novel specificity at the het-c heterokaryon incompatibility locus of Neurospora crassa. Mol Cell Biol 21: 1045-1057.
41. Dix NJ, Webster J. 1995. Fungal Ecology. Chapman & Hall, London.
42. Griffin DH. 1994. Fungal Physiology. Second Edition, Wiley-Liss, New York.
43. Reynolds TB, Fink GR. 2001. Bakers' yeast, a model for fungal biofilm formation. Science 291: 806-807.
44. Schultze K, Schimek C, Wostemeyer J, Burmester A. 2005. Sexuality and parasitism share common regulatory pathways in the fungus Parasitella parasitica. Gene 348: 33-44.
45. Bennett RJ, Uhl MA, Miller MG, Johnson AD. 2003. Identification and characterization of a Candida albicans mating pheromone. Mol Cell Biol 23: 8189-8201.
46. Herranz S, Rodriguez JM, Bussink HJ, Sanchez-Ferrero JC, Arst jr. HN, Penalva MA, Vincent O. 2005. Arrestin-related proteins mediate pH signalling in fungi. Proc Natl Acad Sci USA 102: 12141-12146.
47. Levina NN, Lew RR. 2006. The role of Tipp-localized mitochondria in hyphal growth. Fungal Genet Biol 43: 65-134.
48. Belozerskaya TA. 1998. Cell-to-cell Communication in differentiation of mycelial fungi. Mem Cell Biol 11: 831-840.
49. Bardwell L. 2004. A walk-through of the yeast mating pheromone response pathway. Peptides 25: 1465-76.
50. Dohlman HG. 2002. G proteins and pheromone signalling. Annu Rev Physiol 64: 129-52.
51. Kays AM, Borkovich KA. 2004. Signal transduction pathways mediated by heterotrimeric G proteins. In The mycota III, Brambl R, Marzluf GA (eds) pp 175-207 Berlin: Springer.
52. Hoffman CS. 2005. Except in every detail: comparing and Contrasting G-Protein Signalling in Saccharomyces cerevisiae ans Schizosaccharomyces pombe. Eukaryot Cell 4: 495-503.

53. Beck T, Hall MN. 1999. The TOR signalling pathway controls nuclear localization of nutrient-regulated transcription factors. Nature 402: 689-92.
54. Cutler NS, Pan X, Heitman J, Cardenas MA. 2001 The TOR Signal Transduction Cascade controls cellular differentiation in response to nutrients. Mol Biol Cell 12: 4103-13.
55. Crespo JL, Hall MN. 2002. Elucidating TOR signalling and rapamycin action: lessons from saccharomyces cerevisiae. Microbiol Mol Biol Rev 66: 579-591.
56. Rohde JR, Cardenas ME. 2004. Nutrient signalling through TOR kinases controls gene expression and cellular differentiation in fungi. Curr Top Microbiol Immunol 279: 53-72.
57. D'Souza CA, Heitman J 2001 Conserved cAMP signalling cascades regulate fungal development and virulence. FEMS Microbiol Rev 25: 349-364.
58. Mitchell TK, Dean RA 1995 The cAMP-dependent protein kinase catalytic subunit is required for appressorium formation and pathogenesis by the rice blast pathogen Magnaporthe grisea. Plant Cell 7: 1869–1878.
59. Krüger J, Loubradou G, Regenfelder E, Hartmann A, Kahmann R. 1998. Crosstalk between cAMP and pheromone signalling pathways in Ustilago Maydis. Mol Gen Genet 260: 193-198.
60. Hartmann HA, Kahmann R, Bölker M. 1996. The pheromone response factor coordinates filamentous growth and pathogenicity in ustilago maydis. EMBO J 15: 1632-41.
61. Hartmann HA, Krüger J, Lottspeich F, Kahmann R. 1999. Environmental Signals Controlling sexual development of the Corn Smut Fungus Ustilago maydis through the transcriptional Regulator Prf1. Plant Cell 11: 1293-1305.
62. Kellis M, Birren BW, Lander ES. 2004. Proof and evolutionary analysis of ancient genome duplication in the yeast Saccharomyces cerevisiae. Nature 428: 617-24.
63. Kellis M, Patterson N, Endrizzi M, Birren B, Lander ES. 2003. Sequencing and comparism of yeast species to identify genes and regulatory elements. Nature 423: 241-54.

64. Andersson JO. 2006. Convergent Evolution: gene sharing by eukaryotic plant pathogens. Curr Biol DOI: 10.1016/j.cub.2006.08.042
65. Richards TA, Dacks JB, Jenkinson JM, Thornton CR, Talbot NJ. 2006. Evolution of filamentous plant pathogens: gene exchange across eukaryotic kingdoms. Curr Biol 16: 1857-1864.
66. Latijnhouwers M, de Wit PJ, Govers F. 2003. Oomycetes and fungi: similar weaponry to attack plants. Trends Microbiol 11: 462-469.
67. Van Diepeningen AD, Debets AJM, Hoekstra RF. 2006. Dynamics of dsRNA mycoviruses in black aspergillus populations. Fung Gen Biol doi:10.1016/j.fgb.2006.01.014.
68. Whisson SC, Avrova AO, Lavrova O, Pritchard L 2005 Families of short interspersed elements in the genome of the oomycete plant pathogen, Phytophtora infestans. Fungal Genet Biol 42: 351-65.
69. Rooney AP, Ward TJ 2005 Evolution of a large ribosomal RNA multigene family in filamentous fungi: Birth and dead of a concerted evolution paradigm. PNAS 102: 5084-5089.
70. Selker EU, Tountas NA, Cross SH, Margolin BS, Murphy JG, Bird AP, Freitag M 2003 The methylated component of the Neurospora crasse genome. Nature 422: 893-897.
71. Galagan JE, Selker EU 2004 RIP: the evolutionary cost of genome defense. Trends Genet 20: 417-423.
72. Witzany G. 2006b. Plant Communication from Biosemiotic Perspective. Plant Sign Behav 4: 169-178.

3

Coral communication*

Coordination and organization of coral development and growth through different levels of communication processes

Abstract: Scleractinian corals are keystone species of tropical reef ecosystems. Development and growth of these organisms rely on complex and efficient communication patterns. Such patterns are sign-mediated interactions that cannot be reduced to mere exchange of information, but rather require active coordination and organization. Communication processes within and among corals take place at different levels and occur (a) within cells, (b) among cells, (c) between corals of the same species (conspecies), (d) between corals of different species, (e) and among trans-specific organisms that include all phyla of all kingdoms of shallow-water tropical ecosystems.

This article gives an overview of the manifold levels of coral communication and thereby broadens the understanding of these organisms.

Keywords: manifold communication patterns, ecosystem stability, communication interference, coral disease.

---

* co-author: Pierre Madl; Institute of Physics & Biophysics at the University of Salzburg

1. Introduction

Even though tropical coral reefs are embedded in an oligotrophic setting, they are still capable of sustaining a lush environment. The frequently used description of reefs as the "tropical rainforests of the sea" is rooted in structural conditions that can be compared to terrestrial ecosystems of the tropics: i.e., growth rates are in tune with autochthonous species and their associated symbionts, ensuring a stable, long-term equilibrium, similar to that of tropical rainforests.[1] It is not surprising then, that community structures of coral reefs easily cover a developmental time window stretching over 1000s of years. The sudden emergence of modern reefs across the globe as we know them today, initiated about 6000-9000 years ago.[2,3]

Over the last decades, however, the scientific community noticed a drastic increase in coral-associated diseases of tropical reefs.[4] Caribbean reefs are the hardest hit, assigning this region the nickname "coral disease hot spot".[5]

In this review we will demonstrate that corals, being predominantly sessile organisms, are an essential part of the manifold communicative patterns present within these ecosystems. Thus, internal communicative patterns correspond and are co-evolutionary adapted to external signals of non-coral organisms. The coral animal, therefore, must be able to distinguish between self and non-self. With their tolerance limits, these animals process and evaluate information in order to adequately adjust their responses. Equipped with appropriate recognition patterns, and in absence of external interference, coral animals are able to thrive and establish themselves in oligotrophic environment of tropics.

To understand these highly diverse communicative competences we notice that all this is possible only because communication processes within the animal (intraorganismic) and among the same or related species (interorganismic) as well as species of non-related taxa (metaorganismic) are coordinated in an orderly manner, thereby enabling a context-coherent coordination of behavioral patterns. As long as these processes are comprehended and interactions as well as exchange of information are successfully established, the organism will prosper. Once these patterns become corrupted, via naturally or anthropogenically induced alterations, the coral animal faces an interactive uphill struggle. If there is an external meddling in communication processes, as in the case of chronic events, it will endanger the animal's state of health and ultimately induce disease or even death. In order to understand survival strategies of corals better it is necessary to work out the underlying interactional pattern – a very topical issue especially when considering the effects of global climate change.

We recognize that the use of signal molecules go beyond information exchange, but above all requires the formation of various behavioral patterns necessary for successful interactions. The different forms of the symbiotic association, for example with its endosymbionts and on the mucosa of the coral animal – here exosymbionts – emphasize the fact that conduct among symbiotic partners are beneficial for both parties as long as they are characterized by the absence of stress factors. As there are altruistic forms of interactions between coral and symbionts so there are defensive response patterns, which can involve also life and death situations (see below).

In any case, the context of a given situation determines the meaning of the used signs: (a) growth and (b) development are different modes of behavior and need other patterns of signalling than (c) defense or (d) reproductive patterns. Likewise, do (e) mutual symbioses require different forms of coordination than those of (f) commensalism or (e) parasitism? Thus, this systematic approach of coral communication demonstrates, that the meaning (semantics) of signal molecules is context dependent, and helps to understand the full range of sign-mediated interactions of coral life better.

2. Chemical vocabulary

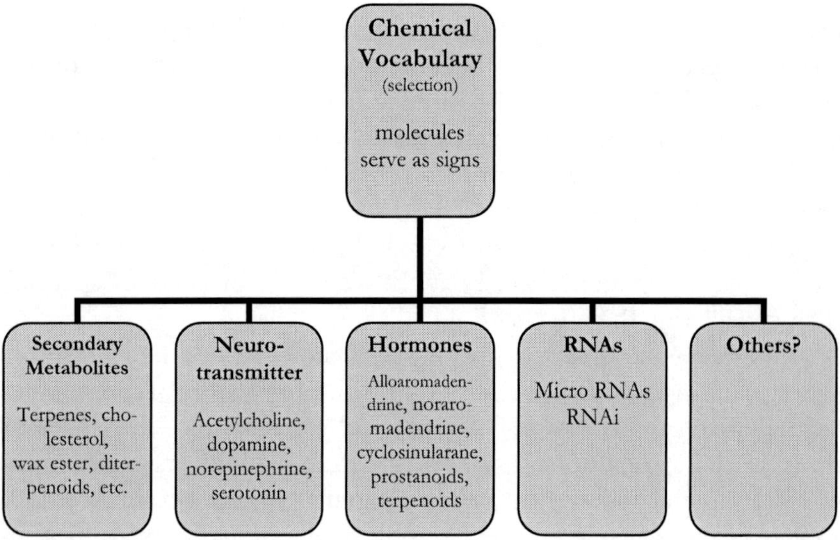

Figure 1: Semiochemical vocabulary in coral communication processes

Communication patterns by signalling molecules within and among corals are amazingly complex. Depending on the developmental stage or other situational contexts, like e.g. the reproductional process, different molecules are used (Fig. 1). Varying behavioral patterns lead to the production of likewise different signals with different functions, such as: antimicrobials, antifungals, corresponding secondary metabolites, and hormones.[6-18]

As it will be shown later, corals possess a broad variety of hormonal substances for different behaviors, e.g. reproduction cycles and defense patterns against opportunistic microbes, carnivores, herbivores and fungal infections.[19,11,8]

3. Interpretation of external impact

Abiotic interferences affect all eukaryotic organisms at any cellular level. Corals are particularly affected, as they are keystone species in tropical reef ecosystems.[20,21] Natural disasters such as earthquakes, cyclones, or tsunamis do change and thereby shape reef ecosystems. Under natural conditions such interference results in change of species composition, favoring pioneering and faster growing species, thereby initiating a new cycle of coral succession. Connell's intermediate-disturbance hypothesis (IDH) gives a good summary on the associated affects of such disturbance.[22-24] Under moderate conditions, abiotic influences enable coral animals to release signalling chemicals via ocean currents.[25,26] Morphogenesis is yet another characteristic of stony corals that is likewise determined by abiotic conditions; i.e., light intensity, swell and surge patterns, oceanic currents.[27-32]

When speaking of anthropogenic influences, eutrophication due to land-based intensive farming or mariculture, excessive fishing practices, and other man-made influences also interfere with coral-ecosystem communication, thereby easily tipping the balance towards a filter-feeding community. High phosphate levels in otherwise oligotrophic waters, for example, rapidly shut down the calcification mechanism (phosphate poisoning). In addition, higher nutrient levels favor (macro)-algae, prevent larval settling, which readily replace the coral-symbiont biotope.[33,24] Once filter feeders substitute the coral carpet and become the dominant organisms, it is almost impossible to re-establish a thriving coral ecosystem within a reasonably short period.[34,35] Closely associated to eutrophication and coastal degradation is the global trend in the decline of mangrove forest cover. Mangroves are an essential part of the wider tropical reef ecosystem, as their numerous prop roots account for a very rich and diverse habitat ranging from algae, sponges, marine invertebrates, nurseries for young shrimp and coral fishes. Habitat degradation for shrimp farms and other coastal utilization not only increases coastal erosion, but likewise interferes with the communication processes within the wider reef ecosystem thereby detrimentally feeding back onto adjacent reefs.[36]

Another man-made disturbance originates from chemical interference; here we refer just to one example which is the pesticide Chlorpyrifos used in golf courses. Coral larvae are extremely sensitive to chemical signals at levels that are well below detectable limits of current human technology. Since crustose coralline algae easily take up this pesticide, coral larvae have lower recruitment rates on substrates exposed to the chemical at concentrations even as low as 5ppb than on untreated controls.[37]

4. Metaorganismic (trans-specific) communication

Both, mechanical and chemical sign-mediated interactions of corals with other genera, families, phyla, as well as with members of other kingdoms are not only essential for their survival, but are the basis of coordination and organization. These interactions cover the entire range – from the mutually supporting over neutral, to harmful behaviors (Fig. 2). The variety of symbiotic communications, for example, requires very different behaviors from the partners involved.[38]

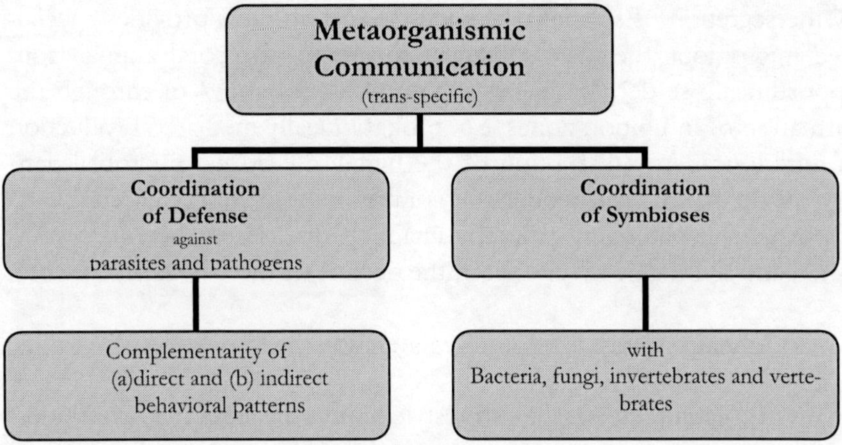

Figure 2: Coral communication with non-coral organisms.

Direct and indirect defense mechanisms are manifold and complement to each other. Corals possess a "non-self" warning system, especially when confronted with opportunistic microbes.[39] Such "microbes" include single-celled autotrophic and heterotrophic prokaryotes, autotrophic and heterotrophic eukaryotes, as well as viruses. In fact, these opportunists can be found in all of the three domains of life on our planet; i.e., Bacteria, Archaea, and Eukarya.[40]

Corals interact with non-related species through their muco-polysaccharide layer (MPSL). This species-specific layer forms a boundary through which dissolved nutrients and gases diffuse. Hence, the mucus is a barrier against opportunistic pathogens and should be considered the primary immune organ of corals. Beneficial prokaryotic residents living on and within the MPSL are yet another shield against opportunistic settlers. There is a large number of mucus-adapted microbes; such as phosphate and nitrogen fixers. However, this is complemented in that corals actively "fish" the surface layer to obtain additional nutrients. In order to do so, corals encourage growth of specific microbes by the secretion of specialized mucus.[41] This, in turn, provides specialized microbiota, the ideal substrate, to protect the coral animal from opportunistic settlers by occupying entry niches and / or through the formation of inhibition zones; e.g. prokaryotically mediated production of antibiotics. Any disruption of the highly diversified microbial density on the MPSL will render the coral holobiont more susceptible to opportunistic pathogens, thereby enabling diseases to become established and ultimately may result in the decline of the whole organism.[39]

4.1 Coordination of defense and regeneration

Being predominantly sessile in lifestyle, corals are equipped with additional defense mechanisms against mechanically induced damage and microbes.[42,43] It turns out that corals can differentiate between the various modes of inflicted tissue damage; i.e., triggered by (a) viruses, (b) bacteria, (c) fungi and (d) invertebrates as well as (e) damage induced by vertebrates. Based on the type of damage, corals secrete different combinations of substances that serve to deal with such lesion.

A localized wound response leads not only to the production of mobile signal molecules, but requires a systematic reaction involving the entire organism – obviously, a mature colony has simply more reserves than a juvenile recruit.[44] Reactions to lesions stimulate the transmigration of specific amoeboid wound cells toward the site of injury. The action of amoebocytes can be considered as first aid measures as they clean up cellular debris. Necrotic tissue is then sloughed off.. Only now will the surrounding tissues stretch to cover as much of the wound as possible. If lesions are too large to be covered by stretching tissues, a much slower process takes over: tissue regrowth over the denuded area. It was observed however, that corals affected by microbial-induced disease, revealed a significant reduction of amoebocytes engaged in tissue repair.[43]

Corals are able to directly and indirectly coordinate their complementing protective measures to varying degrees - based on the intensity of damage inflicted by opportunistic species.[45-47] The exposure to parasites or pathogens stimulates the coral organism to produce a specific array of immune substances.[48-50] In addition, corals also produce enzymes that render their tissues unpalatable to certain predators.[51,52] Contrary to the predictable succession of developmental phases of individual organisms, modular organisms can proliferate at one end while at the other tissues may be already in the phase of decomposition.

Death in such organisms often results from becoming too big or succumbing to disease rather than from programmed senescence. Thus, the body of a modular organism has an age structure – it is composed of young and developing, actively functioning as well as senescent parts.[23,53] The modular structure enables corals to respond adequately to spatial limitations, predators, and unfavorable environmental conditions. Their morphology, in particular branching species, not only provide shelter for juvenile fish species and other invertebrates, but actively benefit the survivorship of the coral: e.g. the crab Trapezia sp. can even deter such vicious corallivorous predators like Acanthaster planci.[54,55]

Although corals are almost at the trophic base of shallow tropical reef biota, they are not passive "prisoners" of their environment. Rather the opposite is the case: being at the base - more appropriate would be at the center of what a reef community is all about they actively shape their environment and thereby provide other organisms from other taxa both habitat and substrate.

However, the fragility of reef-ecosystems becomes evident when external disturbance such as devastating events, e. g. storms or changes in water quality, result in large deviations from normal modal flux rates that can lead to rapid reef decline. When limited in duration - as briefly mentioned with Connell's IDH - such events can stimulate another cycle of reef accretion.[34] In healthy reef ecosystems, this balance of production and bioerosion of reef biomass and calcium-carbonate substrate is balanced with net accumulation.(see Appendix). It regulates and influences ecosystem function of growth and decay.[56,57] However, under the influence of global climate change, non-linear effects of long-lasting adverse conditions can easily lead to a situation where a 'catch-' or 'keep up' reef can easily precipitate into a 'give-up' state, where "permanent" decline sets in.[58]

4.2 Communicative coordination of Symbioses

Generally, there are multiple symbiotic interactions, which can be clearly differentiated into mutualistic, aggressive and defensive properties.[59,60] Due to the gradual increase in global sea-surfaces temperatures, communication processes between endosymbionts and coral hosts are increasingly disturbed. In such cases, the coral's ability to neutralize endosymbiotic production of radicals is compromised.[61]

Hence thermal stress combined with high irradiance pushes the host into the distress phase which leads to ever more frequent expulsion of endosymbiotic algae as temperature extremes exceed thermal threshold levels.[62,63] During extended bleaching events coral communities of entire reef sections lose their ability to regenerate and ultimately fade out, giving rise to a completely altered ecosystem.[64,65]

However, there are some clades of Symbiodinum that are better adapted to the higher temperatures.[66] Under certain circumstances – as claimed by the adaptive bleaching hypothesis (ABH) – corals are able to swap less temperature tolerant clades with better-adapted ones,[67,68] thereby enabling corals to partly regain their vitality.[65,69] The most marked differences among clades of Symbiodinium can be found between corals of the Caribbean and those in the Indo-Pacific region while lesser differences exist among clades along a depth gradient within a given region.[70]

Tropical reef corals arc ecologically important examples of mutually helpful associations, whose success depend on the intrinsically interwoven webs of life. This network of complex communicative interactions shapes a sensitive dynamic equilibrium. Yet, as illustrated previously, this balance can far too easily be disturbed by outside influences thereby triggering the collapse of entire reef sections. Besides active conservation efforts, only an exact analysis of the communication patterns of all the partners involved can aid to a better understanding of coral physiology, disease patterns, biotic and abiotic stress-factors.[64]

5. Interorganismic (species-specific and species-related) communication

Hermatypic reef corals are social organisms and with few exceptions, such as some members of Fungiidae, they are predominantly colonial and modular in appearance. This social capacity implies species-specific sign-mediated interaction processes, which enables corals of same or similar species, as well as distant relatives to coordinate their behavior. This coordination is a most obvious during reproduction.

While some corals reproduce sexually, which requires synchronization of opposite sexes, such as in cases of mass spawning others do so asexually via budding or fragmentation. Most corals employ both modes of reproduction.[71] Gametogenesis is a time and resource-dependent process, hence, either case depends on favorable environmental condition and developmental stage. Laboratory breeding experiments with some Acropora species have shown that pre-mating barriers are not strong. This indicates that chemical messengers are almost absent or not recognized in some of these species to prevent merger of gametes from different species. The effect of such hybridisation can range from sterile individuals to introgression, speciation and even extinction.[72] Hence, it comes of no surprise that species involved in mass spawning overcome this limitation by expelling their gametes at precisely different time windows of the lunar cycle coupled with solar insulation, and/or sea surface temperatures.[25,26]

Evolutionary processes are also induced via hybridization.[73,74] In situ observations along the GBR revealed that in a single night up to 150 species of the highly cross-fertile genus Acropora spawn within hours of each other. High cross-fertilization rates were documented in vivo, while molecular tree topologies confirmed non-monophyletic patterns. These bore little similarity to cladistic analysis based on skeletal morphology or to the fossil record, leading to the conclusion that hybridization does contribute to the enormous success of these species.[75,76] Indeed, hybridization and polyploidy are recognized not only in other coral genera but also among a wide variety of marine invertebrates.[77] Likewise, recombinational crosses between directly and indirectly developing species of sea urchin have resulted in novel gene expression[78] and novel ontogenetic pathways,[79] leading to the creation of new larval morphologies.[75] According to Williamson,[80] interphyletic hybridization of marine invertebrates must be a major transfer route of marine larval forms between taxa through a wide variety of hybridizations.[75]

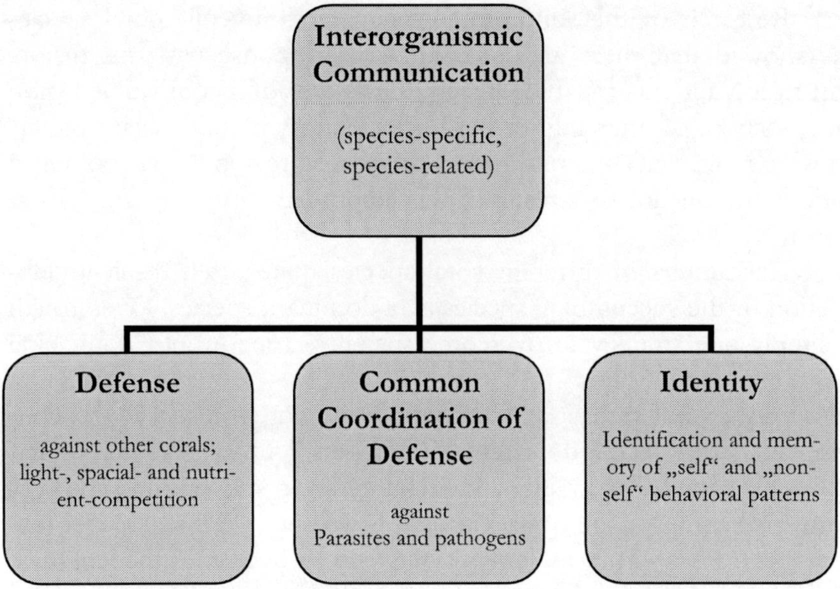

Figure 3: Examples for interorganismic communication.

Corals can differentiate between "self" and "non-self".[49,81] In order to protect its own growth range against proliferating opportunists, defensive measures are taken against "non-self" tissues (Fig. 3). Allelopathic reactions, such as the production of chemicals to signal its presence and to limit excessive proliferation of neighboring coral species, occur in very complex ways and in various combinations and gradations.[82] There are some completely different and complementing defense mechanisms, e.g. escape in size, aggressive behavior, allelochemicals and aggregation.[83] Aggressive and defensive behavioral patterns reciprocally depend on the extent of physical contact.[84,85] Such responses can be quite broad, in that they can involve complete rejection of other coral colonies (use of sweeper and or stinging tentacles) or in the opposite case to result in complete merger of both colonies.[86,87,45]

Research of this kind employing various juvenile coral species has showed that there are three kinds of response patterns: fusion, non-fusion and incompatible fusion. In the case of incompatible fusion the junction of merging tissues lacks endosymbiotic algae. Slowly growing polyps characterize such an interfacial region. Over prolonged periods of time and as a result of this incompatibility, a skeletal barrier is formed.[88,89]

Encounters of different coral species quite often result in subduction of the succumbing species by a dominant species. Overgrowth is simply one strategy to overcome spatial restriction in a flourishing reef community.[88,90]

Spatial and nutritional competition among corals and the concomitant stress factor do affect their fitness.[91] Once resource limitations and habitat constraints are encountered, corals can become very aggressive and even kill competitors. It is well-known that sweeper tentacles, extrusions from the polyp's gut – up to five times the length of the polyp's tentacle and equipped with batteries of nematocysts – literally digest competitors away. These extended mesenterial filaments are part of the macroscopic defense abilities.[92] In addition some corals generate considerable amount of mucus, loaded with nematocysts, which spreads out and over into the nearby environment to harm neighboring colonies. Prolonged mucus production can significantly damage and even kill affected colonies. Others do so via the excretion of chemical poisons into the adjacent water body, while yet others again secrete substances, which render larval settlement of potential competitors unfavourable.[93-99]

Once a coral detects the presence of a nearby species with its ability for non-self recognition, it reacts with the production of finely gradated antibacterial/cytotoxic substances that will be used against potential intruders, or even against related coral species. This is an acquired response pattern and it becomes entrained as a result of continuous stimulation. Corals are able to learn, in that they compare a given stimulation pattern with bodily stored stimulation patterns of the past.[47,50]

Hence, corals are capable of differentiating between tissues and chemicals of kin species from those of "non-related" species. This is essential in order to avoid repercussions that negatively affect the survival rate of nearby individuals of the same species. Fitted with such tools of discrimination, corals are even able to differentiate the sex of their opponents.[100]

Along with their symbiotic partners, each colony must have some kind of sphere of individuality in order to survive and prosper. Once intruders compromise these preconditions, substances are produced and released into the water body that hamper growth and proliferation of nearby competitors.[6,101,11,12,14,7] For the same purpose, even sturdy coral species can produce and release fast-acting antibacterial agents into the environment.[99]

6. Intraorganismic communication

As we will see, intraorganismic communication processes are sign-mediated interactions within cells (intracellular) and between other cells of the same organisms (intercellular). Both communication processes are of crucial importance for the coordination of growth and development. Similar to multicellular organisms the exchange of information must be assured both on a local scale as well as on a spatially larger scale.

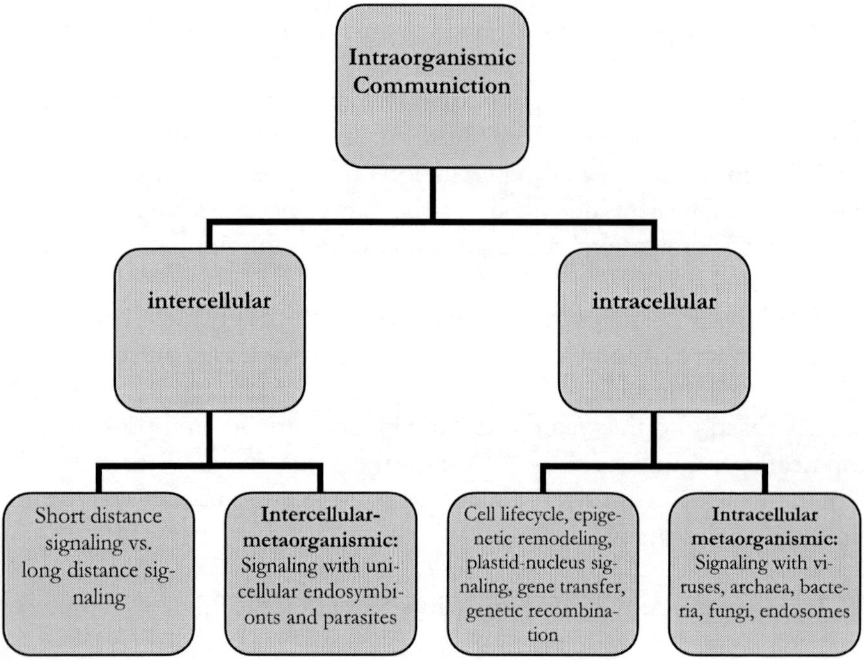

Figure 4: Levels of intraorganismic communication

Only by responding as a whole and in a coherent fashion can coral organisms react to the corresponding developmental challenges and physiological influences.

6.1 Intercellular communication

Intercellular communication serves to communicate events within coral tissues or compartments to remotely located cells or tissues (Fig. 4). An injured coral for example organizes an integrated molecular, biochemical and cell-biological response. This includes also immunological reactions.[47,48,50]

The status information about the coral's current condition is constantly monitored. However, this kind of information is to some extent suppressed when the organism undergoes periods of growth. Doing so enables expression of pre-processed intermediate steps necessary to accommodate such phase transitions. Embryonic development in particular and its subsequent transition into the larval stage as well as the metamorphosis to a juvenile polyp require finely tuned coordinations of growth and development. Special signalling pathways initiate these steps. Neuropeptides for example, are hormone-like substances that coordinate metamorphosis.[102]

Acquisition of environmental parameters is part of the coral's life cycle. This kind of information is even passed on to later generations. This pattern becomes most evident during asexual reproduction when offsprings respond by establishing an extensive carpet of juvenile thickets, while others show preferences for high irradiance, and yet others again cope well with shadier conditions.[84]

6.2 Intracellular communication

Intracellular merger of symbiotic dinoflagellates with marine cnidarians is the most important prerequisite to ensure a highly productive and diverse reef-ecosystem.[103,104] To assert that the effects of symbiosis are long-lasting, this endocytotic process is coordinated by the ApRab5 gene. However, if expression of this gene is disturbed, it leads to sudden separation and expulsion of the symbionts.[105,106]

Thus, successful intracellular communication (Fig.4) between the symbiogenetically assimilated unicellular eukaryotic cells must take place. It makes sure that external messages are transformed and forwarded to the endsoymbionts. This information epigenetically influences the genetic expression of algal DNA. This, in turn, triggers a particular genetic reply, which leads to the production of signal molecules and generates an adequate response behavior.[107]

It seems quite obvious that viroids, viruses and bacteria do interfere in this kind of intracellular communication. Under extreme circumstances, this interference can disturb or even cause the collapse of the coral host – such as in the case of induced tissue bleaching by Vibrio shiloi in Oculina patagonica or V.coralliilyticus in Pocillopora damicornis.[108,109] Indeed most Symbiodinium species are infected with icosahedral double-stranded DNA- viruses. Under normal conditions they replicate without harming the host. Nonetheless, their latent virulence turns lytic (lethal to Symbiodinium) once water temperatures rise, thereby forcing the coral to expel their death endosymbionts, which also leads to tissue bleaching.[110] Microbial interference on the other hand is reciprocal; it enables incorporation of specific genetic features into the intruder's genome as well as to export microbial data sets into that of the host organism. It is very likely that the capability to incorporate different traits into each other is a key principle of symbiogenetic processes.[111]

On a macroscopic scale, corals possess a decentralized neural network.[112] Alignment of this network is never static, but implies neuronal-like plasticity. According to the animal's past experiences, it is the capacity to modify (increase or decrease) the magnitude of their connections. Both memory and functions and long-lasting neuronal plasticity require new RNAs for the appropriate protein synthesis. Such neuronal plasticity implies that signals are relayed via the synapse to the nucleus. Therein, these signals are converted in order to evoke a change in gene transcription (Fig. 4). Only then can the resulting changes (RNAs, proteins) be converted and sent back to the synapse to enable long-lasting change.[113-115] It will be quite interesting to observe investigative efforts that tackle such decoding functions responsible for intracellular communication processes also within corals.

Variation in pigmentation among corals is primarily the result of few genes and their associated proteins of the endosymbionts. It is worth noting that the various color patterns are the result of phenotypic plasticity rather than species diversity as it was previously thought.[116]

The simple organismic structure of diverse groups of Cnidaria phylum sets them almost into the "archaic" section of the animal line as they both share most characteristics with a common ancestor as well as those of modern animals of higher taxa; i. e., large-scale coral-algal-sponge bioherms emerged some 450 million years ago.[2] It should be maintained that evolution, growth and development of the most important coral endosymbionts, i. e., members of the dinoflagellate genus Symbiodinium, assign it a similar age. In this respect and according to the emergence of the first cnidarian precursors, the coral animal can be regarded as a "window to the past".[117] Despite their simplicity, these animals have nearly as many genes as higher animal taxa, including vertebrates. Genome analysis has shown that most genes in Cnidaria can also be found in higher animals. This suggests that selective gene loss, or restriction of gene expression, rather than massive gene enrichment must be the main trigger for the emergence and evolution of higher animal taxa. This also implies that a selective decrease of the genes has contributed to the sheer variety of animal species and paved the way for differentiation of the currently known diversity in families, genera and species.[118]

Discussion

There is compelling evidence that evolution, growth and development of scleractinian corals largely depend on successful meta-, inter- and intraorganismic communication processes. Indeed, it is not the individual coral organism that accounts for prosperous long-term establishment within the wider coral reef ecosystem, but rather it is the relationship and communication processes with its surrounding life-world that enables the coral to survive in the long run. Once these processes are successfully established, coordination and organization within the coral animal (intraorganismic) and with other corals can take place in a controlled manner (Fig. 5). These processes enable corals to proliferate along with other organisms within the tropical reef ecosystem.

Communication processes should not be considered a mere exchange of information, rather they can be substantiated as diverse and manifold sign-mediated patterns of interaction. The use of such signs is subject to semiotic rules.[119] The basic rules are quite reliable and conservative.

Proliferation of corals depends on successful communication, which means the communication processes may even fail. The response of a particular interaction can be misleading, e. g. interaction between the coral and its endo- as well as exosymbionts pretend to be mutualistic, only to draw a temporary advantage from a given interaction and/or even to damage the exploited organism substantially once a shift in environmental conditions has taken place. It is obvious that this cannot be the general form of communication. If all symbionts would behave in this way, no individual species could eventually survive. In the majority of cases, there must be efficient and successful interactions that are beneficial or at least not harmful for all the participants involved. Anything else leads to unsuccessful communication, disease, and decay of the coral organism. Hence, large-scale inefficiency in communication inevitably leads to long-lasting decline of the entire reef ecosystem. From this point of view we suggest that a beneficial behavior should exist among corals that serves to warn neighboring organisms of the same species, as it has already been observed in many other sessile taxa.[120]

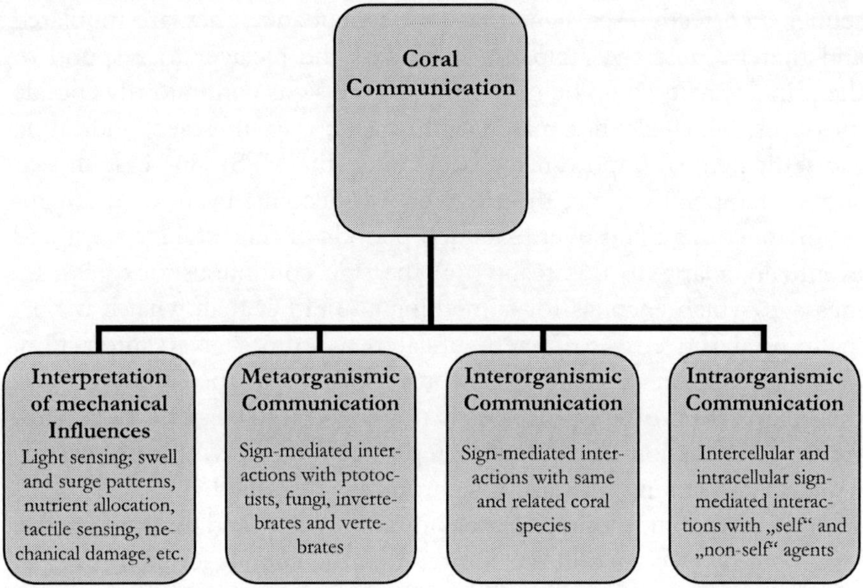

Figure 5: Coral communication: Sign-mediated interactions which each obey three levels of semiotic rules (syntactic, pragmatic, semantic).

Successful communication processes are guaranteed once all participants obey the semiotic rules in signal production, combination as well as transduction and interpretation-induced response behavior. However, this does not imply that these rules are void of errors, distortions, lack completeness, or are absent altogether or in yet other situations even regenerated de-novo. Nor does it mean that rules are always respected. In distinction to natural laws, the rules of sign usage in principle are subject to change. Violation of this kind leads to distorted messages, which induce missing or even erratic response behaviors. There are various ways in which a message can be misinterpreted: (a) the sender incorrectly uses signs that lack redundancy, they are mutilated and thereby miss the recipient; it hinders the receiver to respond to them in a way it would be necessary; (b) the signs continuously encode messages, which do not match with reality; e. g. messages indicating the settlement of opportunistic bacteria on the MPSL, increase in sea-surface temperature, etc; the receiver will become aware of the inappropriate content and over extended periods of time will ignore it and eventually adapts to this reality; (c) the sign continuously expresses a message, which encodes for something else rather than what it is normally used for; e. g. to benefit unilaterally from a given interaction. However and on the basis of chronic violation of the semiotic rules, the organization of life processes is no longer possible; i. e., communicative coordination of evolution, reproduction, growth and development, in and among organisms is chronically disturbed.

If communication processes on meta-, inter- and intraorganismic levels can be discerned in the future, then the various levels of rules in which signs are used can be differentiated. It is obvious that the "molecular syntax"[121] of intracellular sign usage is different when compared to that of intercellular usage. Likewise do syntactic rules, which determine combinations of signalling molecules, they differ in species-specific interactions from those used in trans-specific interactions. Embedded in the ecological contexts and according to the behavioral contexts (pragmatic), different combinations are used. That is why structurally identical messages attain a partially different meaning when used in another context.[122,123] In the future, the integration of this viewpoint could help to decipher coral-specific meanings of semiochemical messages in their entirety. The more we understand these communication processes and the meaning of the used signs, the deeper our understanding of the entire symbiology of corals will be.

Appendix

Bioerosion: Due to the intense competition and predation within a reef ecosystem, selection pressure and niche specialization have led to the selection and evolution of cryptic lifestyles. Many of the secretive species are without toxins, armature, spines, and thick shells. Thus, a huge eroding pressure by numerous corallivorous, endo- and epilithic species from all taxa – ranging from microbial endoliths to grazing vertebrates – does occur even in undisturbed tropical reef ecosystems. The term bioerosion was coined by Neumann[124] to denote the physico-chemical action of various organisms degrading the calcareous substrates. Bioerosion releases calcium-carbonate into the environment, which is then re-used by other accreting organisms. However, this fragile balance between reef accretion and reef erosion is easily upset by abiotic or biotic interference, thereby further enhancing bioerosive processes. A high level of interpretation processes is thereby integrated, which regard both abiotic sensing, meta- and interorganismic exudates as well as their associated response behavior. It has been suggested that their numbers and combined mass of bioeroders equal or exceed that of the surface biota.[56,34]

On the other hand, bioerosion provides innumerable small organisms (protozoans, flatworms, copepods etc.) as well as larger ones like feather stars, crabs etc. Countless fish also use corals to seek shelter (e.g. schools of juveniles) and venture out from branching corals whenever danger has subsided. In a healthy, actively accreting reef, bioeroders contribute enormously to the diversity of habitats within the reef structure. Bioeroders are important in sculpturing coral reef growth and in producing sediments (rubble, sand, silt, and clay). However, if reef growth slows in response to natural or anthropogenic environmental stress factors, the dynamic balance is disturbed, inter- and transspecific communication distorted, which over the long run, leads to excessive bioerosion and ultimately triggers the decline of entire reefs.

References

1. Hubbell SP. 1997. A unified theory of biogeography and relative species abundance and its application to tropical rain forests and coral reefs. Coral Reefs 16:9-21.
2. Hallock P. 1997. Reefs and Reef Limestones in Earth History. In: Birkeland C (ed) Life and Death of Coral Reefs. Chapman and Hall, London, pp 13-42.
3. Pandolfi J. 2002). Coral community dynamics at multiple scales. Coral Reefs 21:13-23.
4. Sutherland K, Porter JW, Torres C. 2004. Disease and immunity in Caribbean and Indo-Pacific zooxanthellate corals. Mar Ecol-Prog Ser 266:273–302.
5. Weil E. 2004. Coral Reef Diseases in the Wider Caribbean; In: Rosenberg E, Loya Y (eds) Coral Health and Disease. Springer Verlag, Berlin-Heidelberg, pp 35-59.
6. Kim K 1994. Antimicrobial activity in gorgonian corals (Coelenterata, Octocorallia). Coral Reefs 13:75-80.
7. Slattery M, James A, McClintock B, Heine JN. 1995. Chemical defenses in Antarctic soft corals: evidence for antifouling compounds. J Exp Mar Biol Ecol 190:61-77.
8. Slattery M, Hines GA, Starmer J, Paul VJ. 1999. Chemical signals in gametogenesis, spawning, and larval settlement and defense of the soft coral Sinularia polydactyla. Coral Reefs 18:75-84.
9. Ramesh P, Venkateswarlu Y. 1999. Novel steroid constituents of the soft coral Sinularia dissecta. Steroids 64:785-9.
10. Iwashima M, Nara K, Iguchi K 2000. New marine steroids, yonarasterols, isolated from the okinawan soft coral, Clavularia viridis. Steroids 65:130-137.
11. Kim K, Kim PD, Alker AP, Harvell CD. 2000 a. Chemical resistance of gorgonian corals against fungal infections. Mar Biol 137:393-401.
12. Kim K, Harvell CD, Kim PD, Smith GW, Merkel SM. 2000 b. Fungal disease resistance of Caribbean sea fan corals (Gorgonia spp.). Mar Biol 136:259-267.

13. Yasumoto M, Mada K, Ooi T, Kusumi T. 2000. New terpenoid components from the volatile oils of the soft corals Clavularia viridis and Sarcophyton acutangulum. J Nat Prod 63:1534-1536.
14. Roussis V, Chinou IB, Tsitsimpikou C, Vagias C, Petrakis PV. 2001. Antibacterial secondary metabolites Antibacterial activity of volatile secondary metabolites from Caribbean soft corals of the genus Gorgonia. Flavour Frag J 16:364-366.
15. Twan WH, Hwang JS, Chang CF. 2003. Sex Steroids in Scleratinian Coral, Euphyllia ancora: Implication in Mass Spawning. Biol Reprod 68:2255-2260.
16. Watanabe K, Sekine M, Iguchi K. 2003. Isolation of three marine prostanoids, possible biosynthetic intermediates for clavulones, from the Okinawan soft coral Clavularia viridis. Pharm Bull 51:909-913.
17. Iguchi K, Kukaya T, Yasumoto A, Watanabe K. 2004. New marine sesquiterpenoids and diterpenoids from the Okinawan soft coral Clavularia koellikeri. J Nat Prod 67:577-583.
18. Zhang GW, Ma XQ, Kurihara H, Zhang XC, Yao XS, Su JY, Zeng LM. 2005. New hemiketal steroid from the soft coral Cladiella sp. Org Lett 7:991-994.
19. Hay ME, Fenical W, Gustafson K. 1987. Chemical defense against diverse coral-reef herbivores. Ecology 68:1581-1591.
20. Bak RPM, Termaat RM, Dekker R. 1982. Complexity of coral interactions; influence of time, location of interaction and epifauna. Mar Biol 69:215-222.
21. McClanahan TR, Maina J. 2003. Response of coral assemblages to the interaction between natural temperature variation and rare warm-water events. Ecosystems 6:551-563.
22. Connell JH. 1978. Diversity in tropical rainforests and coral reefs. Science 1999:1302-1310.
23. Begon M, Harper JL Townsend CR. 1996. Ecology, Individuals, Populations and Communities, 3rd Edition, Blackwell Science, Oxford.
24. Loya Y. 2004. The Coral Reefs of Eilat – Past, Present and Future. In: Rosenberg E, Loya Y (eds) Coral Health and Disease. Springer Verlag, Berlin-Heidelberg, pp 2-28.

25. Harrison PL, Babcock RC, Bull GD, Oliver JK, Wallace CC, Willis BL. 1984. Mass spawning in tropical reef corals. Science 223:1186-1189.
26. Penland L, Loulechad J, Idip D, van Woesik R. 2004. Coral spawning in the western Pacific Ocean is related to solar insolation: evidence of multiple spawning events in Palau. Coral Reefs 23:133-140.
27. Geistner J. 1977. The Influence of Wave Exposure on the Ecological Zonation of Caribbean Coral Reefs; Proc 3rd Int Coral Reef Symp, Miami, 2:23-29.
28. Horiguchi T, Kawai H, Kubota M, Takahashi T, Watanabe M. 1999. Phototactic responses of four marine dinoflagellates with different types of eyespot and chloroplast. Phycol Res 47:101-107.
29. Gleason DF, Edmunds PJ, Gates RD. 2005. Ultraviolet radiation effects on the behavior and recruitment of larvae from the reef coral Porites astreoides. Mar Biol 148:503-512.
30. Stambler N, Dubinsky Z. 2005. Corals as light collectors: an integrating sphere approach. Coral Reefs 24:1-9.
31. Vargas-Angel B, Colley SB, Hoke SM, Thomas JD. 2006. The reproductive seasonality and gametogenic cycle of Acropora cervicornis off Broward County, Florida, USA. Coral Reefs 25:110-122.
32. Veron JEN, Stafford-Smith M. 2000. Corals of the World. Volume 3, Australian Institute of Marine Science, Townsville.
33. Tanner JE. 1995. Competition between scleractinian corals and macroalgae: An experimental investigation of coral growth, survival and reproduction. J Exp Mar Biol Ecol 190:151-168.
34. Hatcher BG. 1997. Organic Production and Decomposition. In: Life and Death of Coral Reefs. Birkeland C (ed) Chapman and Hall, London, pp 140-174.
35. Elmqvist T, Folke C, Nyström M, Peterson G, Bengtsson J, Walker B; Norberg J. 2003. Response diversity, ecosystem change, and resilience. Front Ecol Environ 1:488–494.
36. Mumby PJ, Edwards AJ, Arias-Gonzales JE, Lindeman KC, Blackwell PG, Gall A, Gorczynska MI, Harborne AR, Pescod CL, Renken H, Wabnitz CC, Lewellyn G. 2004. Mangroves enhance the biomass of coral reef fish communities in the Caribbean. Nature 427:533-536.

37. Richmond RH. 1997. Reproduction and Recruitment in Corals: Critical Links in the Persistence of Reefs; In: Life and Death of Coral Reefs. Birkeland C (ed) Chapman and Hall, London, pp 175-197.
38. Weis VM, Reynolds WS, deBoer MD, Krupp DA. 2001. Host-symbiont specificity during onset of symbiosis between the dinoflagellates Symbiodinium spp. and planula larvae of the scleractinian coral Fungia scutaria. Coral Reefs 20:301-308.
39. Rohwer F, Kelley S. 2004. Culture-Independent Analyses of Coral-Associated Microbes. In: Rosenberg E, Loya Y (eds) Coral Health and Disease. Springer Verlag, Berlin-Heidelberg, pp 265-274.
40. Sherr E, Sherr B. 2000. Marine Microbes – an Overview. In: Kirchman DL, Williams PJB (eds) Microbial Ecology of the Oceans. Willey-Liss Inc., New York, pp 13-38.
41. Kushmaro A, Kramarsky-Winter E. 2004. Bacteria as a source of Coral Nutrition; In: Rosenberg E, Loya Y (eds) Coral Health and Disease. Springer Verlag, Berlin-Heidelberg, pp 231-238.
42. Gunthorpe L, Cameron AM. 1990. Toxic exudates from the hard coral Goniopora tenuidens. Toxicon 28:1347-1350.
43. Kramarsky-Winter E. 2004. What Can Regeneration Process Tell Us About Coral Disease? In: In: Rosenberg E, Loya Y (eds) Coral Health and Disease. Springer Verlag, Berlin-Heidelberg, pp 216-227.
44. Alker AP, Kim K, Dube DH, Harvell CD. 2004. Localized induction of a generalized response against multiple biotic agents in Caribbean sea fans. Coral Reefs 23:397-405.
45. Chadwick-Furman N, Rinkevich B. 1994. A complex allorecognition system in a reef-building coral: delayed responses, reversals and non-transitive hierarchies. Coral Reefs 13:57-63.
46. Koh EGL. 1997. Do Scleractinian Corals Engage in Chemical Warfare Against Microbes? J Chem Ecol 23:379-398.
47. Rinkevich B. 2004. Allorecognition and xenorecognition in reef corals: a decade of interactions. Hydrobiologia 530/531:443-450.
48. Bigger CH, Olano CT. 1993. Alloimmune cellular responses of the gorgonian coral Swiftia exserta, J Immunol 150:134.

49. Rinkevich B, Frank U, Bak RPM, Müller WEG. 1994. Alloimmune responses between Acropora hemprichi conspecifics: nontransitive patterns of overgrowth and delayed cytotoxicity. Mar Biol 118:731-737.
50. Hildemann WH, Raison RL, Cheung G, Hull CJ, Akaka L, Okamoto J. 1977. Immunological specificity and memory in a scleractinian coral. Nature 270:219-223.
51. Lindquist N, Hay ME. 1996. Palatability and chemical defense of marine invertebrate larvae. Ecol Monogr 66:431–450.
52. Kelman D, Benayahu Y, Kashman Y. 1999. Chemical defence of the soft coral Parerythropodium fulvum fulvum (Forskal) in the Red Sea against generalist reef fish. J Exp Mar Biol Ecol 238: 127–137.
53. Vytopil E, Willis B. 2001. Epifaunal community structure in Acropora spp. (Scleractinia) on the Great Barrier Reef: implications of coral morphology and habitat complexity. Coral Reefs 20:281-288.
54. Gosliner TM, Behrens DW, Williams GC. 1996. Coral Reef Animals of the Indo-Pacific – Animal life from Africa to Hawaii exclusive of the vertebrates . Sea Challengers, Monterey.
55. Pratchett MS. 2001. Influence of coral symbionts on feeding preferences of crown-of-thorns starfish Acanthaster planci in the western Pacific. Mar Ecol-Prog Ser 214:111–119.
56. Glynn PW. 1997. Bioerosion and Coral-Reef Growth: A Dynamic Balance. In: Birkeland C. (ed) Life and Death of Coral Reefs. Chapman and Hall, London, pp 68-95.
57. Paulay G. 1997. Diversity and Distribution of Reef Organisms; In: Life and Death of Coral Reefs. Birkeland C (ed) Chapman and Hall, London, pp 299-344.
58. Adey WH. 1978. Coral Reef Morphogenesis: A Multidimensional Model. Science 202: 831-837.
59. Van Veghel MLJ, Cleary DFR, Bak RPM. 1996. Interspecific interactions and competitive ability of the polymorphic reef-building coral Montastrea annularis. B Mar Sci 58:792-803.
60. Hay ME. 1997. The ecology and evolution of seaweed-herbivore interactions on coral reefs. Coral Reefs 16:67-76.
61. Lesser MP. 2004. Experimental biology of coral reef ecosystems. J Exp Mar Biol Ecol 300:217– 252.

62. Hoegh-Guldberg O. 1999. Climate change, coral bleaching and the future of the world's coral reefs. Mar Freshwater Res 50:839-866.
63. Hoegh-Guldberg O. 2004. Coral Reefs and Projections of Future Change. In: Rosenberg E, Loya Y (eds) Coral Health and Disease. Springer Verlag, Berlin-Heidelberg, pp 463-481.
64. Edmunds PJ, Gates RD. 2003. Has Coral Bleaching Delayed Our Understanding of Fundamental Aspects of Coral–Dinoflagellate Symbioses? BioScience 53:976-980.
65. Rowan R. 2004. Coral bleaching: Thermal adaptation in reef coral symbionts. Nature 430:742.
66. Mostafavi PG. 2006. Predominance of clade D Symbiodinium in coral reefs off the coast of Iran. Symbiosis 13:4.
67. Buddemeier R, Baker A, Fautin DG, Jacobs JR. 2004. The adaptive Hypothesis of Bleaching. In: Rosenberg E, Loya Y (eds) Coral Health and Disease. Springer, Berlin-Heidelberg, pp 427-440.
68. Hoegh-Guldberg O, Jones RJ, Ward S, Loh WK. 2002. Is coral bleaching really adaptive? Nature 415:601-602.
69. Baker AC, Starger CJ, McClanahan TR, Glynn PW. 2004. Corals' adaptive response to climate change. Nature 430:741.
70. Toller WW, Rowan R, Knowlton N. 2001. Zooxanthellae of Montastraea annularis Species Complex: Patterns of Distrubution of Four Taxa of Symbiodinium on Different Reefs and Across Depths. Biol Bull 201:348-359.
71. Miller KJ, Ayre DJ. 2004. The role of sexual and asexual reproduction in structuring high latitude populations of the reef coral Pocillopora damicornis. Heredity 92:557-568.
72. Wallace C. 1999. Staghorn Corals of the World – A Revision of the genus Acropora. CSIRO Publishing, Collingwood.
73. Márquez LM, van Oppen MJH, Willis BL, Miller DJ. 2002). Sympatric populations of the highly cross-fertile coral species Acropora hyacinthus and Acropora cytherea are genetically distinct. Proc. Natl. Acad. Sci. USA 269:1289–1294.
74. Miller DJ, van Oppen MJH. 2003. A 'fair go' for coral hybridisation. Mol Ecol 12:805–807.

75. Ryan FP. 2006. Genomic creativity and natural selection: a modern synthesis. Biol J Linn Soc 88:655-672.
76. Van Oppen JH, McDonald BJ, Willis B, Miller DJ. 2001. The evolutionary history of the coral genus Acropora (Scleractinia, Cnidaria) based on a mitochondrial and a nuclear marker: reticulation, incomplete lineage sorting, or morphological convergence? Mol Biol Evol 18:1315-1329
77. Pfenninger M, Reinhardt F, Streit B. 2002). Evidence for cryptic hybridisation between different evolutionary lineages of invasive clam genus Corbicula (Veneroida, Bivalvia). J Evol Biol 15:818–829.
78. Nielsen MG, Wilson KA, Raff EC, Raff RA. 2000. Novel gene expression patterns in hybrid embryos between species with different modes of development. Evol Dev 2:133–144.
79. Raff EC, Popod EM, Sly BJ, Turner FR, Villinski JT, Raff RA. 1999. A novel ontogenetic pathway in hybrid embryos between species with different modes of development. Development 126:1937–1945.
80. Williamson DI. 2003. The origins of larvae. Kluwer Academic Publishers, Dordrecht.
81. Rinkevich B, Sakai K. 2001. Interspecific interactions among species of the coral genus Porites from Okinawa. Jpn J Zool 104:91-97.
82. Yamazato K, Yeemin T. 1986. Preliminary study on the inter- and intraspecific interactions among corals of Khang Khao Island, the Sichang Islands, Gulf of Thailand. Galaxea 5:163-174.
83. Bruno JF, Witman JD. 1996. Defense mechanisms of scleractinian cup corals against overgrowth by colonial invertebrates. J Exp Mar Biol Ecol 207:229-241.
84. Bak RPM, Engel MS. 1979. Distribution, abundance and survival of juvenile hermatypic corals (Scleractinia) and the importance of life history strategies in the parent coral community; Mar Biol 54:341-352.
85. Ferriz-Dominguez N, Horta-Puga G. 2001. Short-term aggressive behavior in scleractinian corals from La Blan-quilla reef, Veracruz Reef System. Rev Biol Trop 49:67-75.
86. Connell JH. 1976. Competitive interactions and the species diversity of corals. In: Mackie GO (ed) Coelenterate Ecology and Behavior . Plenum Press, New York, pp 51-58.

87. Cope M. 1982). Interspecific coral interactions in Hong Kong. Proc 4th Int Coral Reef Symp, Manila, 2:357-362.
88. Veron JEN. 1986. Corals of Australia and the Indo-Pacific. University of Hawaii Press. Honululu.
89. Hidaka M, Yurugi K, Sunagawa S, Kinzie RA. 1997. Contact reactions between young colonies of the coral Pocillopora damicornis. Coral Reefs 16:13-20.
90. Frank U, Rinkevich B. 2001. Alloimmune memory is absent in the Red Sea hydrocoral Millepora dichotoma; J Exp Zool 291:25-39.
91. Tanner JE. 1997. Interspecific competition reduces fitness in scleractinian corals. J Exp Mar Biol Ecol 214:19-34.
92. Barnes R, Hughes R. 1999. An Introduction to Marine Ecology; 3rd ed. Blackwell Science, Oxford.
93. Lang JC. 1970. Inter-specific aggression within the scleractinian reef corals, Ph.D. thesis, Yale Univ., New Haven, Connecticut, p 80.
94. Lang JC. 1971. Interspecific aggression by scleractinian corals; 1. The rediscovery of Scolymia cubensis (Milne-Edwards and Haime). B Mar Sci 21:952-959.
95. Lang JC. 1973. Interspecific aggression by scleractinian corals; 2. Why the race is not only to the swift. B Mar Sci 23:260-279.
96. Logan A. 1984. Interspecific aggression in hermatypic corals from Bermuda. Coral Reefs 3:131-138.
97. Logan A. 1986. Aggressive behavior in reef corals: A strategy for survival. Sea Frontiers 32:347-351.
98. Lang JC, Chornesky EA. 1990. Competition between scleractinian reef corals- A review of mechanisms and effects. In: Dubinsky Z (ed) Coral Reefs. Ecosystems of the World 25. Elsevier, Amsterdam, pp 209-252.
99. Geffen Y, Rosenberg E. 2005. Stress-induced rapid release of antibacterials by scleractinian corals. Mar Biol 146:931-935.
100. Ates R. 1989. Aggressive behavior in corals. Fama 12:104-112.
101. Wilsanand V, Wagh AB, Bapuji M. 1999. Antibacterial activities of anthozoan corals on some marine microfoulers. Microbios 99:137-145.
102. Iwao K, Fujisawa T, Hatta M. 2002. A cnidarian neuropeptide of the GLW-amide family induces metamorphosis of reef-building corals in the genus Acropora. Coral Reefs 21:127-129.

103. Santos SR, Taylor DJ, Kinzie A, Hidaka M, Sakai K, Coffroth MA. 2002. Molecular phylogeny of symbiotic dinoflagellates inferred from partial chloroplast large subunit (23S)-rDNA sequences. Mol Phylogenet Evol 23:97-111.
104. Takabayashia M, Santos CR, Cook CB. 2004. Mitochondrial DNA Phylogeny of the Symbiotic Dinoflagellates (Symbiodinium, Dinophyta). J Phycol 40:160-164.
105. Chen MC, Cheng YM, Hong MC, Fang LS. 2004. Molecular cloning of Rab5 (ApRab5 in Aiptasia pulchella and its retention in phagosomes harboring live zooxanthellae. Bioche Bioph Res Co 324:1024-1033.
106. Chen MC, Hong MC, Huang YS, Liu MC, Cheng YM, Fang LS. 2005. ApRab11, a cnidarian homologue of the recycling regulatory protein Rab11, is involved in the establishment and maintenance of the Aiptasia-Symbiodinium endosymbiosis. Biochem Bioph Res Co 338:1607-1616.
107. Chen CA, Wallace CC, Yu JK, Wei NV. 2000. Strategies for Amplification by Polymerase Chain Reaction of the Complete Sequence of the Gene Encoding Nuclear Large Subunit Ribosomal RNA in Corals. Mar Biotechnol 2:558-570.
108. Rosenberg E. 2004. The Bacterial Disease Hypothesis of Coral Bleaching; In: Rosenberg E, Loya Y (eds) Coral Health and Disease. Springer Verlag, Berlin-Heidelberg, pp 445-457.
109. Rozenblat YBH, Rosenberg E. 2004. Temperature-Regulated Bleaching and Tissue Lysis of Pocillopora damicornis by the Novel Pathogen Vibrio coralliilyticus. In: Rosenberg E, Loya Y (eds) Coral Health and Disease. Springer Verlag, Berlin-Heidelberg, pp301-321.
110. Villarreal LP. 2005. Viruses and the Evolution of Life. ASM Press, Washington D.C.
111. Shackelton LA, Holmes EC. 2004. The evolution of large DNA viruses: combining genomic information of viruses and their hosts. Trends Microbiol 12:458-465.
112. Westfall JA, Sayyar KL. 1997. Ultrastructure of Neurons and Synapses in the Tentacle Epidermis of the Sea An-emone Calliactis parasitica. J Morphol 232:207–216.

113. Moccia R, Chen D, Lylcs V, Kapuya E, Yaping E, Kalachikov S, Spahn CMT, Frank J, Kande ER, Barad M, Martin KC. 2003. An unbiased cDNA library prepared from isolated Aplysia sensory neuron processes is enriched for cytoskeletal and translational mRNAs. J Neurosci 23:9409-9417.
114. Martin KC. 2004. Local Protein Synthesis during Axon Guidance and Synaptic Plasticity. Curr Opin Neurobiol 14:305-310.
115. Thompson KR, Otis KO, Chen DY, Zhao Y, O'Dell TJ, Martin KC. 2004. Synapse to nucleus signalling during long-term synaptic plasticity: a role for the classical active nuclear import pathway. Neuron 44:997-1009.
116. Kelmanson IV, Matz MV. 2003. Molecular Basis and Evolutionary Origins of Color Diversity in Great Star Coral Montastraea cavernosa (Scleractinia: Faviida). Mol Biol Evol 20:1125-1133.
117. Margulis L, Schwartz KV. 1988. Five Kingdoms. W. H. Freeman and Company, New York
118. Miller DJ. 2005. Lessons from coral genomics - perspectives on the development and evolution of higher animals. Biologisches Kolloquium der Christian-Albrechts-Universität, Kiel.
119. Witzany G. 2006. The *Logos* of the *Bios* 1. Contributions to the Foundation of a three-leveled Biosemiotics. Umweb, Helsinki.
120. Kobayashi Y, Yamamura N, Sabelis W. 2006. Evolution of talking plants in tritrophic context: Conditions for uninfested plants to attract predators prior to herbivore attack. J Theor Biol doi:10.1016/j.jtbi.2006.05.026
121. Witzany G. 1995. From the "logic of the molecular syntax" to molecular pragmatism; Evol Cog 1:148-168.
122. Witzany G. 2005. Natural history of life: History of communication logics and dynamics. S.E.E.D. J 5:27-55.
123. Witzany G. 2005. From Biosphere to Semiosphere to Social Lifeworld; Biology as an Understanding Social Science. tripleC 3:51-74.
124. Neuman AC. 1966. Observations on coastal erosion in Bermuda and measurements of the boring rate of the sponge Cliona lampa. Limnol Oceanogr 11:92–108.

4

Communicative Competences of Bacteria

Abstract: Bacteria communicate and therefore are able to organize and coordinate their behavior like a multicellular organism. Communication processes are sign-mediated interactions. Signs are chemical molecules which serve as signals both within and between prokaryotic organisms. Bacteria are symbiotic organisms in the whole range from mutualism to parasitism. They may be beneficial for their (eukaryotic) hosts, without them hosts would not survive. Others are neutral, i.e., do not harm the host and many of them also cause diseases with sometimes epidemic character and often letal consequences. Bacteria are important hosts themselves for multi-viral colonization and virally determined gene-word order.

This review will show examples of the different communicative competences of bacteria on all levels. It means the communication processes within the organism (intraorganismic), between species specific and species-related organisms (interorganismic) and between bacteria and non-bacterial organisms (metaorganismic) as well as their capability to interpret abiotic influences.

Keywords: multi-level communication, sign-mediated interactions, genome editing.

1. Introduction

Although bacteria are relatively simple organisms they represent one of the main successful stories of evolution. They started at the early beginning of life similarely to archaea which represent a different organismic kingdom. Bacteria are found in all ecological niches, share a common flux of their gene-pool with a high rate of gene order recombination for adaptational purposes.[1] More than in any other organismic kingdom it is accepted to speak about the languages of bacteria.[2-4]

'Quorum sensing' is the term to describe sign-mediated interactions in which chemical molecules are produced and secreted by bacteria.[5-7] If these molecules reach a critical concentration in a special ratio to population density they are identified by the bacterial community as indicators to generate an appropriate response behavor.[8] These molecules trigger the expression of a great variety of gene transcriptions. Many bacteria use multiple quorum sensing codes, each may be modulated by post-transcriptional or other regulatory engineering.[9]

There are also communication processes between different species of bacteria and between bacteria and non-bacterial life, such as eukaryotic hosts.[10] Beneath the semiochemicals (Gr.: Semeion = sign) necessary for developmental processes of bacterial communities such as division, sporulation, and synthesis of secondary metabolites, there are physical contact-mediated behavioral patterns being important in biofilm organisation.[11-13] Also abiotic influences serve as signs which indicate specific nutrients or other environmental circumstances like hydrodynamic changes.

Since communities of bacterial species, which are able to coordinate their behaviour, have advantages over single bacteria organisms it is much more common, that the evolutionary drive went into a rising communicative complexity.[14] We should not forget that in comparison to the first 2 billion years of life on earth with closed prokaryotic symbiology the rise and growth of the eukaryotic super-kingdom was a crucial advantage for bacterial lifestyle to colonize vertical hosts with their great motility resources.

We can differentiate three classes of molecules signalling for different purposes, i.e., signalling within the organism to coordinate gene expressions to generate adequate response behaviour, signalling between the same or related and different species. With a limited number of molecules and a limited number of combinatorial rules they generate quite different sign-mediated interactions for different purposes.

As in every sign-mediated interaction sign-users share a common set of syntactic rules, i.e., how signs may be combined; of pragmatic rules which determine a great variety of interactional contents, e.g. development, growth, mating, virulence, attack and defense. The situational context of these complex interactional processes determines the meaning of signs, i.e., semantics of signals. Independent of organismic complexity the complementarity of these three levels of semiotic rules we can in principle identify in every sign-mediated interaction within and between organisms.[15,16]

This leads to the generation of intra- and intercellular processes which enable bacterial communities to generate memory which may be inheritable but can alter epigenetically, i.e., different reading/meaning patterns of the same genetic dataset with differences on a phenotypic level without altering the genetic dataset.

Although there is a great awareness of communiative competences of bacteria the terms of description of signs, signals, intra- and interspecific sign-mediated interactions are rarely unique. I suppose 3 semiotic levels wherein signalling molecules are embedded: (i) the syntactic level, i.e., their combinatorial patterns, their (ii) pragmatic rules which determine the interactional contents of behavior and (iii) the semantic level of rules which determine the meaning function of the chemical and/or physical signals.

This review will show that there is a way for uniqueness in terms of description which is oriented by biosemiotics which investigates primarily sign-processes within and between organisms of all organismic kingdoms.[16]

2. Interpretation and coordination

Because bacteria have profound effects on human health, agriculture, industry and other ecospheres they are target to multiply drugs which fight them.[17] They develop drug resistance by coordination of special defense behavior called biofilm organisation, which depends on the competence of bacteria for chemotaxis. Biofilm organisation is a special kind of coordination with high density of physical contact and contact specific signalling. If bacteria realize a critical mass via quorum sensing they build up a high density of social body by moving their flagellas which may resist even strong antibiotics.[18] Biofilms are constructed on both abiotic surfaces as well as biotic ones, e.g. on stones in rivers and other aqueous surfaces, or e.g. in the respiratory track of animals. Nutrient availability regulates also the structure of biofilm organisation[19] as hydrodynamic forces.[20] Interestingly, it has been found that biofilm organisation is linked with coordinated DNA release which is integrated in it.[21]

3. Semio-Chemical vocabulary of bacterial communication

In receiving signals from same or related species or non-bacterial organisms the signalling molecules bind to specialized sensor proteins which function as receptors. They transmit the message to an intracellular regulator.[22,23] Then the signal molecule transits the cell membrane through diffusion or by specific transport pathways. Inside the cell the signalling molecule in most cases binds to a cytoplasmic target protein. It may happen that a diffusible molecule is chemically engineered to an active signal after entering the target cell.[23] Organization of cellular production of response molecules leads to signal-dependent transcription control of DNA.

The semiochemical vocabulary used by bacteria is of great variety, especially because some signalling molecules are multiply re-usable components.[24] Today three kinds of communicative goals are distinguished: (A) reciprocal communication, active sign-mediated interactions which is beneficial for both interacting parts; (B) messages which are produced as response on a triggering event which may be an indicator for a receiver which has not been specially targeted by the producer. A coincidental event which is neutral – except of the energy costs of production – to the producer but beneficial for the receiver; (C) signalling to manipulate the receiver, i.e., to cause a response behavior which is one-sided beneficial to the producer and harms the receivers[23] often in that they behave against their normal interests.[25] These three communicative goals are also detectable in eukaryotic communication processes.

The three classes of intra-, inter- and metaorganismic communication enable bacteria to generate and coordinate different behavioral patterns: Self and non-self identification, i.e., recognition and identification of other colonies and measurement of their size, pheromone-based courtship for mating, alteration of colony structure in formatting of fruiting bodies, initiation of developmental and growth processes e.g. sporulation.

Acyl homoserine lactones (AHLs) and linear oligopeptides are used as signs in diverse processes. Cyclized oligopeptides function as virulence genes. g-Butryolactones (GBLs) are used as antibiotics and in sporulation processes. Furanosyl diester (AI-2) is used in diverse processes[26] and in luminescence. *cis*-11-Methyl-2-dodecenoic acid (DSF) serves in virulence and pigmentation. 4-Hydroxy-2-alkyl quinolines (PQS, HAQs) are important in the whole regulation processes and for virulence as do palmic acid methyl esters (PAME). Putrescine is important in swarming motility like biofilm organisation. A-signal is used in early developmental processes and aggregation. C-signal is a cell surface-associated protein and serves to coordinate motility and the developmental process of fruiting body building. Cyclic dipeptide is a secondary metabolite.[23]

Gram-negative bacteria use homoserine lactones (LuxR/LuxI) as signs in communication processes,[27] whereas gram-positive bacteria use oligopeptides in quorum sensing communication. As in all organisms non-coding RNAs are important in higher order regulatory pathways. Small RNAs and microRNAs are used by bacteria to regulate special genetic expression patterns which play an important role as appropriate response behavior to stress or nutrient availability[28], e.g. in controlling the quorum sensing pathways.[29] Bacteria have to distinguish between species-specific signalling, and signalling which is able to modulate behaviors interspecifically.[4] With these communicative competences they are able to coordinate species-specific behavorial patterns as well as to coordinate behaviors between diverse species.

4. Metaorganismic communication

Starting with beneficial symbioses between bacteria and plants we refer to the complex communication networks between soil bacteria, mychorrizal fungi and plant roots. Mychorrizal fungi secrete molecules in the surrounding environment which serves as nutrients for soil bacteria and triggers their activation to degrade special nutrients which are then available for mychorrizal fungi.

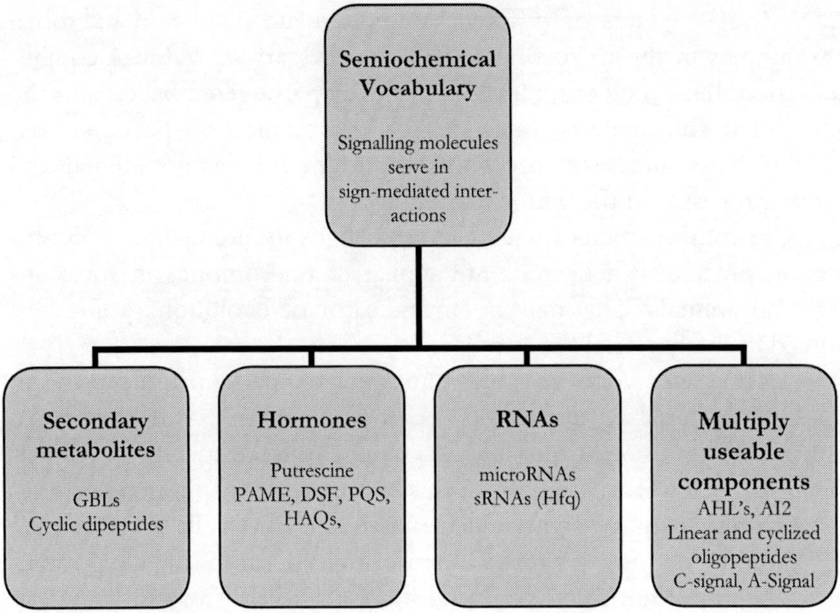

Figure 1: Examples of semiochemicals used in bacterial communication processes

Their hyphal growth serves as developmental and growth area of plant roots. Plant roots themselves depend on nutrients which are prepared by the mychorrizal fungi. They can also mimic bacterial signalling molecules either to trigger bacterial production of special molecules or to disturb bacterial communication pathways.[30-32,8]

Rhizobia bacteria are integrated into plant cells by phagocytosis when they interact symbiotically with the plant roots.[33,34] In other cases where rizhobia fail to fix nitrogen inside the root nodules because they are cheating, plants are sanctioning these rhizobia[35] and prevent their spread in order to stabilize mutualistic symbioses with bacterial colonies.[25] Root exudates of different kinds regulate plant and microbial communities in the rhizosphere. This is necessary to stabilize equilibrium and inhibit the continuity of attacks by pathogenic bacteria in the soil.[36,37] The full range of trans-specific communication processes between bacteria and plant roots are important for developmental and growth processes in the whole plant kingdom.[6,38,32]

Chemical molecules which serve as signs in intercellular communication processes of bacteria are similar to pheromones in social insects and animals. This may be an indicator of evolutionary lineages, evolved in the bacterial "chatter".[39]

Marine eukaryotes are able to mimic bacterial quorum sensing to inhibit bacterial successful communication.[40] Interbacterial communication uses hormone-like signalling to trace specific host locations such as intestinal habitat. In this specialized ecosphere a bacteria-host communication occurs which means the host cells and bacterial cells share a common meaning function for the same signalling molecules.[41]The important role of bacteria for the whole history of evolution is documented in their endosymbiotic lifestyle as potential candidates for symbiogenesis[42-45] (documented in the origin of eukaryotic endosomes like mitochondria[46]).[15,47] The interactions may be pericellular colonization events but also an intracellular lifestyle. These different symbiotic interactions range from acquisition of novel genetic material to a dramatic reduction in size and content connected with gene loss.[48] Successful living processes of higher eukaryotes would not be possible without beneficial symbiosis with bacteria. For example, the cell mass of an adult human resembles 20 % of human origin and up to 80 % of exogenic settlers[49] most of them bacteria.

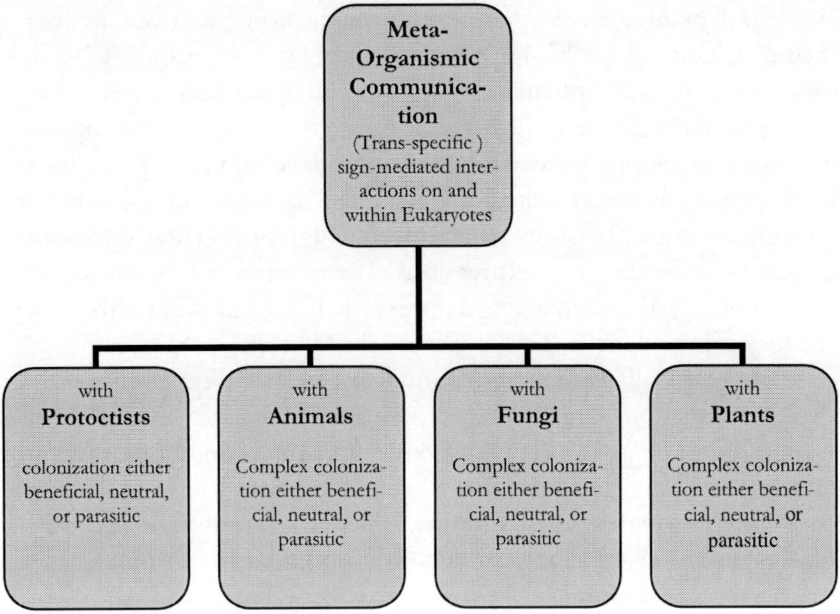

Figure 2: Metaorganismic (trans-specific) communication levels in bacterial communication

5. Interorganismic communication

For a long time it was assumed that bacteria live predominantly as monads. Meanwhile it has been recognized that this is the rare exception.[50,51] Bacterial colonies live in nearly all cases not alone but in coexistence with other bacterial species self-coordinated by a diversity of sign-mediated interactions.[52,53] In all ecological in vivo situations bacteria use intraspecific and interspecific signalling.[25] This implies also a broad variety of conflicts within and between species.[54] The mutual, neutral and manipulative aims of communication processes as mentioned in Chapter 3 of this review are special kinds of response behavior to certain degrees of beneficial up to conflictual relationships.[25]

Dependent on the availability of nutrients some bacteria suppress normal cell development which leads to the development of a different cell type which is better suited for adequate response behavior in this situational context. It means that different environmental conditions can lead to different gene expressions. If the same colony is exposed several times to these changing contexts it has been shown that they react more immediately. This indicates that bacterial communities are able to develop collective memory and learn from the experience.[55,52]

This functions are similar to neuronal networks in higher eukaryotes. In the case of changing environmental conditions the suppression of cell division may lead to cell elongation which enables cell colony to change modus of motility. This is an important feature of socio-bacterial behavior, e.g. swarming coordination and organisation for surface colonization.[56]

Some authors documented altruistic strategies in mixed colony formations which seem to be an advantage of microcolonies. Altruistic behavioral strategies enable to strengthen self-identity and a sustainable equilibrium in multilevel colonized ecological niches.[57,58]

Interestingly, bacteria use a common contextual interpretation of incoming signals by each member of the colony. The response behavior is appropriate to the majority of "votes"[56] which represents the decision, i.e., how each bacterial colony member measures the incoming signals of all others.

The identification of non-self species is a competence possible through species-specific and group-specific quorum sensing and is coherent with the assumption that smaller groups of the same bacterial species are able to build a kind of quorum sensing-"dialects". These is important in high density of coexistent bacterial life habitats to prevent confusion and enable more complex coordination.[59] Interestingly, the prokaryotic cell-cell communication has structural analogues to cross-kingdom signalling between bacteria and fungi.[60]

Bacteria decide in special cases to form fruiting bodies of different types and shape for sporulation. This enables bacterial communities to a more efficient dissemination of the spores. The fruiting body building is governed by pragmatic rules with different roles to different sub-groups of bacterial communities for coordination.[61] Some have to serve for motility to density, followed by direction decision and decision of cell types, cell growth and developmental stages in all different steps until fruiting body is ready for sporulation event. Without communicative hierarchical organization this would not be possible. If communication is disturbed, body building is not insured, so bacterial communities have developed special strategies to single out the so called cheaters;[55] it does not follow the rules of behavioral coordination.

One of the most interesting and most investigated phenomenon of bacterial communication is the symbiology of multiple colonies coexistent in the human oral cavity (Figure 4).[62-64] Bacteria on human teeth and oral mucosa establish a homeostasis of pathogenic and mutualistic bacteria by a complex system of sign-mediated interactions both species-specific and trans-specific. The dental plaque in the oral cavity of humans is a unique habitat which is not found in any other species. The homeostasis is not static but the result of a dynamic relationship between different species-colonies dependent on intervals of daily hygiene. There are approximately 500 different interacting species.[65-67] Each member of these communities must be capable of self and non-self distinction, and of distinguishing between species-specific signalling and trans-specific signalling. As community they must be able to measure their own colony size and the size of the other colonies and molecules which have the same chemical structure but are not part of a biotic message.[62,63]

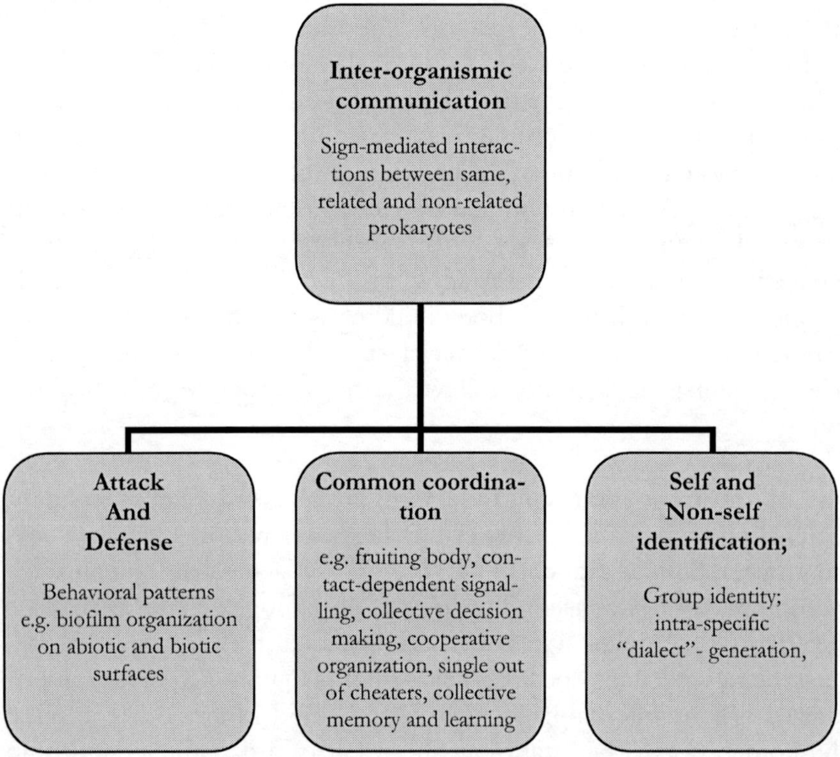

Figure 3: Species-specific, species-related and non-related prokaryotic communication

Special communication patterns with detailed hierarchical steps of signal production and transmission include (i) metabolite exchange, (ii) cell-cell recognition, (iii) genetic exchange, (iv) host signal recognition and signal recognition of same or related species. Because of the high number of competing and cooperating species there is a special short- and long-term community architecture established. If the communication on the intra-, inter- and metaorganismic level is successful, i.e., the signal transmission and reception enables colonies to live in a dynamic homeostasis, then the human oral cavity will miss cavity diseases.

6. Intraorganismic communication

Interestingly, prokaryotic gene order is not as conserved as the protein sequences. Only some higher order regulations (operons) which code for physically interacting proteins are found in almost all bacterial and archaeal genomes. In difference to eukaryotic organisms and few archaea where the protein vocabulary underlies higher order regulatory functions encoded in the non-coding DNA; bacteria lack such gene order conservation.

This indicates high dynamics of new gene orders as documented in the horizontal gene transfer events with its intensive intragenomic recombination.[68,69] This exchange of whole genes or gene-blocks enable bacterial lifestyles to combine several bacterial competences, i.e., phenotypes. The transformation process includes the release of naked DNA, followed by the uptake and recombination, i.e., the integration, with 17 steps identified until now (see fig. 3, chapter 7 in this book).[70] So we can recognize the outcomes of a diversity of mobile DNA-contents,[71] not a mass of individualized genetic texts, but a bacterial gene pool as text repertoire which is available for each individual bacteria and the resource for bacterial genome innovation and evolution.[72]

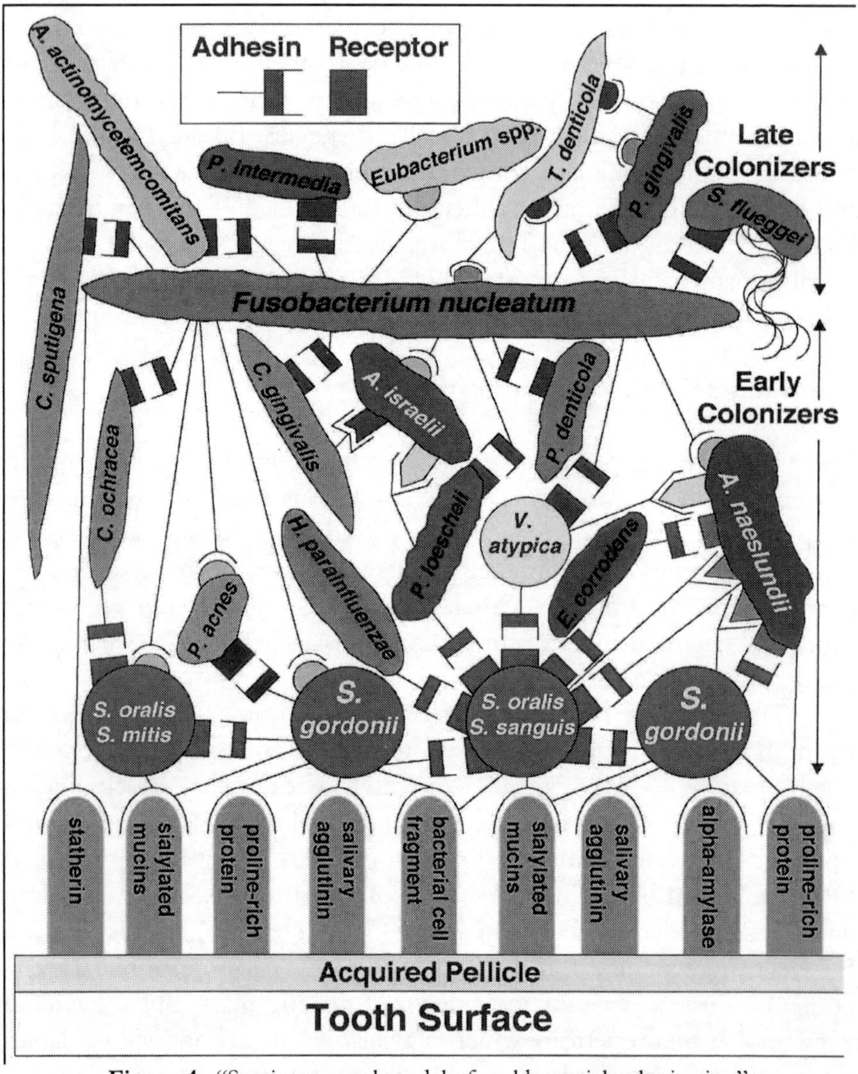

Figure 4: "Spatiotemporal model of oral bacterial colonization" (from: Kolenbrander et al. 2002)

Horizontal gene transfer is a main resource for integrating newly evolved genes into existing genomes and does not need slow steps of chance mutations to alter the genomes but accelerated genome innovations in both bacteria and archaea.[73-75] Important in this context of genomic innovation is not only the sequence acquisition but also the contextualization which means also their loss as another functional feature.[76] As it seems today the phylogeny of microbial species is not a tree of life, but an evolutionary network or a ring of life, mediated by genetic exchange, i.e., acquisition and loss of genetic data-sets.[77,78]

6.1. Intracellular communication

Signal-dependent transcription regulation of the DNA serves for a great variety of response behavior. One of the most interesting phenomena is the fact that in the early 2 billion years of life on the planet Earth the immense density of bacterial life was not an event of the mass of individual organisms but their commonly shared gene-pool which was in constant flux as we know today through investigations on horizontal gene transfer. It means that the evolution of bacteria was not a random event of chance mutations and their selection but transfer of whole genes and gene-blocks representing real phenotypes which were transferred. This led to different combinatorial patterns of genetic encoded phenotypes and the raise of bacterial diversity. It also enabled bacterial pathogens to optimize there disease causing coordination and was therefore a target to special kind of drug developments for medical purposes.[79]

For a long time it has been proposed that tubulin plays an important role in cytoskeletal functions of eukaryotes whereas prokaryotes lack this system. Recent research has shown that tubulin is a very ancient system for chromosome segregation also in bacteria which plays important roles in filament formation, movement and orientation.[780-82]

6.2. Bacterial evolution and natural genome editing by viruses

To elucidate communicative competences of bacteria we also have to look on the role of viruses and their relationship to bacteria. Viruses have long been accepted only as disease causing, epidemic phenomenon with lytic and, therefore, extremely dangerous consequences for infected organisms. Meanwhile new research has corrected this picture.[83]

Viruses are part of the living world in most cases integrated in the cytoplasm or in the nucleoplasma of cells without harming the host. Viruses are on the way to represent the best examples for symbiotic relationships, because there is no living being since the beginning of life which has not been colonized by them, in most often cases in form of multiple colonizations.[83] The longest period of these symbiotic relationships during evolutionary history share viruses, archaea and bacteria.

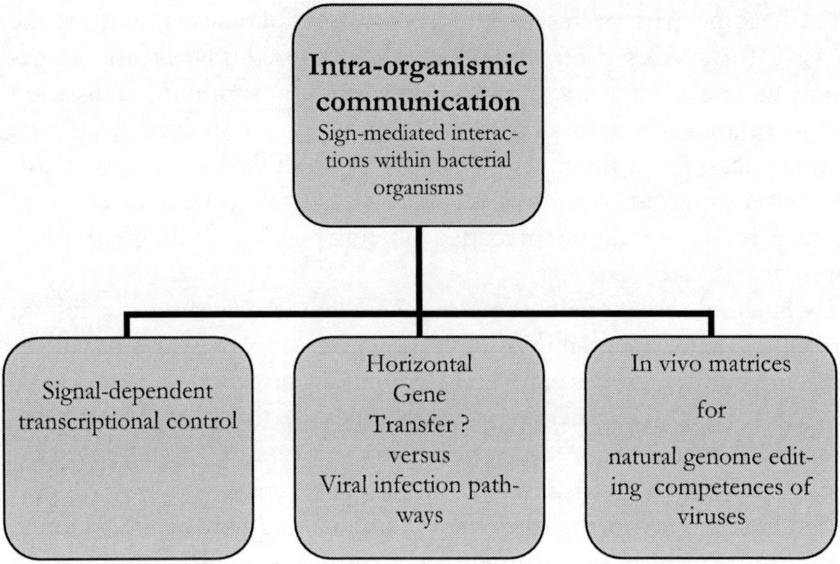

Figure 4: Communication processes within prokaryotes

Because viruses are extremely biosphere-specific, i.e., they adapt to special host tissues, the identification of various forms of, e.g. bacteria is to identify primarily the viruses which colonize them. This is also the concept of "bacteriophages", where bacteria are identified best by defining the viruses which are associated with them.[83] Host identification in this way is a special method called 'phage typing'.

6.2.1 Actute-lytic and temperate-persitent viral life strategies

As mentioned in last years the lytic consequences of viral infection have been a special case if viruses are not able to develop a sessile lifestyle without harming the host. In most cases viruses live within organisms and help to ward off competing parasites from the host, i.e., viruses become part of his evolutionary history. Persistent viruses are decisive for species diversity and host genome editing. Nearly all genome editing competences like expression, transcription, translation and recombination with all its detailed substeps are derived from viral competences. Even the invention of DNA and DNA replication pathways after a period of early RNA world[84-86] seems to be a special viral strategy for conservation of coded phenotypes by warding off RNA parasites.[87,83]

Since observations became more evident, that viruses are able to integrate genetic material into the host genome it was clear that viruses have infection lifestyles and endosymbiotic and even symbiogenetic lifestyles. They give phenotypic capabilities to the host which non-infected hosts from the same species doesn't possess. As endosymbiotic viruses which are dependent on the hosts replication they are part of the host history in that they are inheritable and part of the genomic identity of the host as documented in some several 10 000 infection events in the human genome by endogenous retroviruses.[88]

The two viral lifestyles are not in strict opposition but in most cases are a part of a symbiotic process. It starts with an infection by a virus. In the infected host it comes to an equilibrial status where the immune system does not eliminate the virus but controls its replication without fatal consequences for the host organism. The persistent status lasts during most phases of host life, but may return to the lytic life style if host immune system is under stress.[89] Therefore, the persistence is sometimes called temperate lifestyle. A good example is the persistent virus in all symbiodinium species being the essential endosymbiotic partner for coral animals (see chapter 3 of this book). Coral bleaching as the worldwide phenomenon of coral disease is the consequence of dying of the coral endosymbiont because of global (water)warming. As we know now, dying occurs because the persistent viruses of symbiodinium turns from persistent tolytic lifestyle as reaction on the changing water temperature.[83]

Bacteria may be infected by viruses without being harmed. If infected bacteria meet non-infected it may happen that the non-infected gets lysis, the lysogenic strain does not lyse itself, but is lethal to the non-infected one. The colonized bacteria have a virus-derived molecular genetic identity which has an advantage over the non-infected one through an acquired ability. This lysogenic agent termed prophage has an immunity function for the bacteria which the non-infected bacterium lacks. Prophage is a virus which is integrated into the bacterial host genome. Both the actute lytic phages and the persistent prophages like T4 and lamda are highly abundant in the oceans and in the soil and seem to be the most dynamic life form on the entire planet.[83] Some viruses are not integrated in the host genome but persist as plasmids and replicate independently from the host genome.

If we speak about the relationship of bacteria and viruses we speak about phage ecology. Most prokaryotic viruses are double-stranded DNA viruses with either linear or circular genome morphology and are packaged in an icosahedral capsid. 96% bacterial phages are tailed and only 4% are isometric. In contrast to this there are only 5% of tailed phages in archaea.

Whereas acute viruses in most cases code for their own replication, recombination and repair proteins the persistent phages lack such genes and use the host-cellular replication. This implicates a totally different gene-word order[82] in acute lytic and in persistent phages. This is documented in the very different nucleotide-words (di-, tri- and tetranucleotides). Nucleotide word frequences of actute phages are very dissimilar to that of their hosts while persistent or temperate phages share nucleotide word frequences with the host, it means the molecular syntax[90,91] from actute and persistent phages is constructed totally different according to the different strategies. Different life strategies with different behavioral pathways need a complete different semantic content in the genome.

Because the bacterial cell walls differ substantially among different types of bacteria, viral behaviour should also differ for identification, attachment and penetration. Because of these diverse barriers of bacterial cell walls the prokaryotic viruses do not enter the host cells physically but attach to the cell surface and inject their genomes through contractile tails or pilot-proteins. Also the progeny of the virus has to deal with this barrier.[83]

Bacterial DNA does not have highly stable structures like eukaryotes and can interact with the cellular replication and transcription. In most cases it is circular with a unique origin of replication system. On the contrary to that viral double-stranded DNA is a linear DNA with integrated short terminal repeats. Because bacterial viruses do not use a transport technique like they do in eukaryotes to be transported out of the nucleus; bacterial viruses differ extremely from eukaryotic viruses.

All bacteria have a restriction/modification system which is a connected form of two ancient competing viral competences. Only the descendants of mitochondria lack this restriction/modification system which means, they are not exposed to viral selection occurring in competing viral colonizing events. Maybe, they have transposed their ability to the eukaryotic nucleus which cares in a more efficient way for their cell immunity.[83]

6.2.2 The viral origin of natural genome editing

What we call horizontal gene transfer between bacteria as being responsible for genetic plasticity in prokaryotes may be a capability which is acquired by viral infection pathways. Then viral genetic inventions are transfered to bacteria via lytic and persistent lifestyles of viruses and it is not an exchange phenomenon between bacteria.
Bacteria are the most genetically adaptable organisms with enormous capabilities to react appropriately to extreme changes of their ecologi-

cal habitats. This does not stem from their high reproductive rates but from their high ability to acquire DNA segments which represent complete and complex sets of genes from external sources. These sets are not present in any direct cellular predecessor and derived from viral infections. The infections are not lytic but persistent, i.e., integrated in the host genome. Multiple drug resistance, virulence factors or phage exclusion are three examples of rapid bacterial adaptation which is clearly of a viral origin.[83]

When we look at the whole age of the oceans and the dense abundance of bacterial and viral life in it, then we can say that the possibility of genetic arrangements, rearrangements und exchange, does not need long time distances to create the basics of complexity of life, because the exchange rate is of astronomical order. If we imagine that 1 ml of seawater contains 1 million of bacteria and ten times more of viral sequences, we have to realize that 10^{31} bacteriophages infect 10^{24} bacteria per second.[79] This is an ongoing process. The enormous viral genetic diversity in the oceans seems to have established pathways for the integration of complete and complex genetic data-sets into host genomes, e.g. acquisition of complex new phenotypes via a prophage can include the acquisition of over 100 new genes in a single genome editing event.[83]

Because viruses have induced genomic plasticity of bacteria, bacteria are an ideal global biotic matrix to evolve and develop varieties in genome editing. Bacteria are the smallest living organisms with relatively simple genomic structure where the competitive situation between an abundance of viral infective elements leads to the adaptation of lytic viruses to temperate viruses integrated as plasmids in cytoplasma and even persistent viruses integrated in the host genome. The viral competences could develop in this global bacterial habitat as the bacterial species by their immense genetic flux between viral colonization events and immunity reactions like restriction/modification.

Main genome editing functions like replication, transcription, translation, recombination and all substeps evolved primarely in the competitive situation between viral competences to colonize a host and to ward off competing parasites. This includes that biotic self and non-self identification functions as we know it from diverse immune systems also are of viral origin, i.e., the integration and all genetic/genomic modification steps that what we call natural genome editing are of viral origin. Therefore, the immense importance of horizontal gene transfer for evolution, diversity and competences of bacterial species derived from viral genome editing competences and in most cases was induced by virus-infection.[83] As phylogenetic analyses demonstrate the main protein enzymes for natural genome editing are viral inventions, that are not of a cellular origin.[88,83] Also the origin of eukaryotic nucleus was thought to be an ancient prokaryote but phylogenetic analysis shows that its ancestor was a large DNA virus.[92-94,83]

According to phylogenetic analyses the essential results of genome editing are not emerged randomly through chance mutations and their selection but through real viral competences. This includes replicase, polymerase, integrase, DNA repair, restriction/modification, methylation, bilayer nuclear envelope, eukaryotic nucleus, division of transcription and translation, nuclear pores, tubulin based chromosome duplication, chitin, calcification, linear chromosomes, innate immune system (MHC-Complex, RNAi), adaptive immune system, cartilage, bones, skin, dermal glands for poison, mucus and milk, larvae, egg, placenta, flowering plants, viviparous mammals.[83] Interestingly, the early genetic invention of capsid proteins detected in viruses infecting archaea seems also to be of viral origin and a common ancestry to eukaryotic and bacterial viruses.[95-97]

The virally induced invention of all basic forms of proteins during bacterial evolution corresponds to the crucial step from bacterial analog proteincoding DNA word order to the more sophisticated symbolic ("digital") DNA word order especially of higher eukaryotes. As in every linguistic utterance the complexity of information content raises exponentially with symbolic languages. The symbolic DNA code emerged with repeats, transposons, retroposons and interspersed segments, i.e., transposable and repeated elements, all remnants of viral integration events, which are now identified to be part of a genome editing meta-code.[15] This meta-code in the non-coding DNA has higher order regulatory functions for the same genetic data-set which enables multiple protein meanings and multiple phenotypes to be inherent in the same genetic dataset. For the future this represents the important role of epigenetics and the pathways for inheritance of acquired abilities.

7. Perspectives

The communicative competences of bacteria enable them to develop, organize and coordinate rich social life with a great variety of behavioral patterns which represents one of the most successful life histories of all organismic kingdoms.

Although they exist nearly 4 billion years they still survive, being part of the most dramatic changes in evolutionary history such as DNA invention, invention of nearly all protein types, eukaryotic cells, vertical colonization of all eukaryotes, high adaptability through "horizontal gene transfer" and multi-species colonization of all ecological niches.

They serve as in vivo matrices for natural genome-editing competences of viruses during whole history of life being efficient agents and context building organisms for genomic evolution and genomic evolutionary diversity. "Horizontal gene transfer" seems to be one of the main forces in bacteria evolution because it has been derived from a commonly shared, and transferable gene pool where new genetic inventions have a broad variety to be integrated and recombined. Bacteria, the main sources for symbiotic relationships, without eukaryotes would not have evolved and would not be able to survive. If we agree biology to be a history of symbiology then bacteria are the driving force of it.

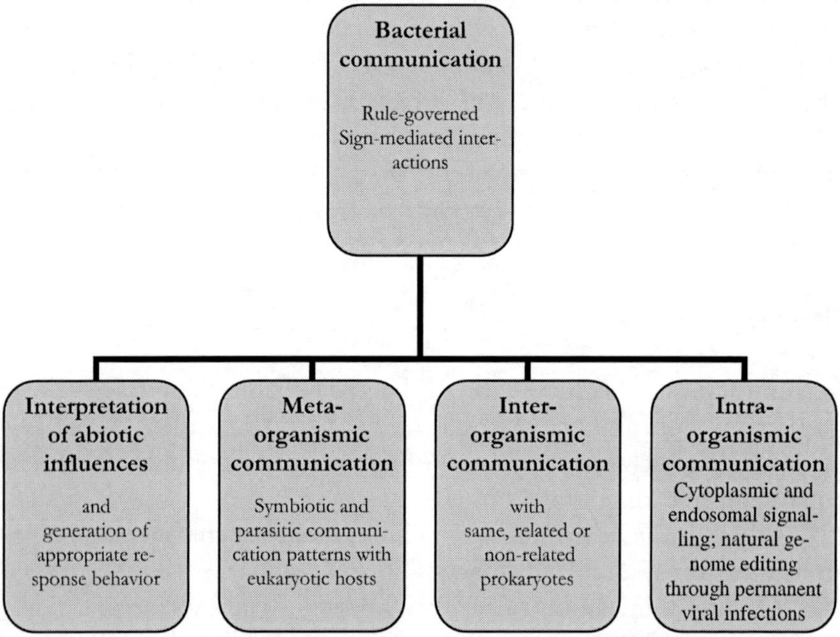

Figure 5: Levels of bacterial communication

Recent research from phylogenetic analyses demonstrates that these competences of bacteria have originated from the capability of viruses for natural genome editing with an astonishing variety of genomic creativity. Bacteria seem to be the optimal biotic matrix for viral induced genetic inventions. Therefore, the communicative competences of bacteria and the genome-editing competences of viruses play a complementary role and are inseperable.

References

1. Pal C, Papp B, Lercher MJ. 2005. Adaptive evolution of bacterial metabolic networks by horizontal gene transfer. Nat Gen 37:1372-1375.
2. Bassler BL. 1999. How bacteria talk to each other: Regulation of gene expression by quorum sensing. Curr Opin Microbiol 2:582-587.
3. Bassler BL. 2002. Small talk: cell-to-cell communication in bacteria. Cell 109:421-424.
4. Schauder S, Bassler BL. 2001. The languages of bacteria. Gen Develop 15:1468-1480.
5. Crespi, BJ. 2001. The evolution of social behaviour in microorganisms. Trends Ecol Evol 16:178-183.
6. Manefield M, Turner SL. 2002. Quorum sensing in context: out of molecular biology and into microbial ecology. Microbiol 148: 3762-3764.
7. Greenberg EP. 2003. Bacterial communication: tiny teamwork. Nature 424:134.
8. Daniels R, Vanderleyden J, Michiels J. 2004. Quorum sensing and swarming migration in bacteria FEMS Microbiol Rev 28:261-289.
9. Loh J, Pierson EA, Pierson LS 3rd, Stacey G, Chatteriee A. 2002. Quorum sensing in plant associated bacteria. Curr Opin Plant Biol. 5:285-290.
10. Konaklieva MI, Plotkin BJ. 2006. Chemical communication – do we have a quorum? Mini Rev Med Chem 6:817-825.
11. Davis DG, Parsek MR, Pearson JP, Iglewski BH, Costerton JW, Greenberg EP 1998. The involvement of cell-to-cell signals in the development of a bacterial biofilm. Science 280:295-298.
12. Fuqua C, Greenberg EP. 2002. Listening in on bacteria: acetylhomoserine lactone signalling. Nat Rev Mol Cell Biol 3:685-695.
13. Voloshin SA, Kaprelyants AS. 2004. Cell-cell interactions in bacterial populations. Biochemistry (mosc) 69:1268-1275.
14. Ben Jacob E. 2003. Bacterial self-organization: co-enhancement of complexification and adaptability in a dynamic environment. Philos Transact A Math Phys Eng Sci 361:1283-1312.

15. Witzany G. 2005. Natural history of life: history of communication logics and dynamics. S.E.E.D. J 5:27-55.
16. Witzany G. 2006. The *Logos* of the *Bios* 1. Contributions to the foundation of a three-leveled biosemiotics. Umweb, Helsinki.
17. Camara M, Williams P, Hardman A. 2002. Controlling infection by tuning in and tuning down the volume of bacterial small-talk. Lancet Infect Dis 2:667-676.
18. Wadhams GH, Armitage JP. 2004. Nat Rev Mol Cell Biol 5:1024-1037.
19. Stanley NR, Lazazzera BA. 2004. Environmental signals and regulatory pathways that influence biofilm formation. Mol Microbiol 52:917-924.
20. Wuertz S, Okabe S, Hausner M. 2004. Microbial communities and their interactions in biofilm systems: an overview. Water Sci Technol 49:327-336.
21. Spoering AL, Gilmore MS. 2006. Quorum sensing and DNA release in bacterial biofilms. Curr Opin Microbiol 9:133-137.
22. Fuqua C, Winans SC, Greenberg EP. 1996. Census and consensus in bacterial ecosystems: the LuxR-LuxI family of quorum-sensing transcriptional regulators. Annu Rev Microbiol 50:727-751.
23. Visick KL, Fuqua C. 2005. Decoding Microbial Chatter: Cell-Cell Communication in Bacteria. J Bacteriol 187:5507-5519.
24. Henke JM, Bassler BL. 2004. Bacterial social engagements Trends Cell Biol 14:648-656.
25. Keller L, Surette MG. 2006. Communication in bacteria: an ecological and evolutionary perspective. Nat Rev Microbiol 4:249-258.
26. Sun J, Daniel R, Wagner-Dobler I, Zeng AP. 2004. Is autoinducer-2 a universal signal for interspecies communication: a comperative genomic and phylogenetic analysis of the synthesis and signal transduction pathways BMC Evol Biol 4:36.
27. Schauder S, Shokat K, Surette MG, Bassler BL. 2001. The LuxS family of bacterial autoinducers: biosynthesis of a novel quorum-sensing signal molecule. Mol Microbiol 41:463-476.
28. Vogel J, Sharma CM 2005. How to find small non-coding RNAs in bacteria. Biol Chem 386: 1219-1238.

29. Lenz DH, Mok KC, Lilley BN, Kulkarni RV, wingreen NS, Bassler BL 2004. the small RNA Chaperone Hfq and multiple small RNAs control quorum sensing in vibrio harveyi and vibrio cholerae. Cell 118: 69-82.
30. Teplitski M, Robinson JB, Bauer WD. 2000. Plants secrete substances that mimic bacterial N acylhomoserine lactone signal activities and affect population density dependent behaviors in associated bacteria. Mol Plant Microbe Interact 13:637-48.
31. Bauer WD, Robinson JB. 2002. Disruption of bacterial quorum sensing by other organisms. Curr Opin Biotechnol 13:234-237
32. Witzany G. 2006a. Plant Communication from Biosemiotic Perspective. Plant Sign Behav 4:169-178.
33. Samaj J, Baluska F, Voigt B, Schlicht M, Volkmann D, Menzel D. 2004. Endcytosis, actin cytoskeleton, and signalling. Plant Physiol 135:1150-1161.
34. Hayashi M. Plastid proteins crucial for symbiotic fungal and bacterial entry into plant roots. Nature 2005; 433:527-31.
35. Kiers ET, Rousseau RA, West SA, Denison RF. 2003. Host sanctions and the legume-Rhizobium mutualism. Nature 425:78-81.
36. Walker TS, Bais HP, Grotewold E, Vivanco JM. 2003. Root exudation and rhizosphere biology.Plant Physiol 132:44-51.
37. Bais HP, Park SW, Weir TL, Callaway RM, Vivanco JM 2003. How plants communicate using the underground information superhighway. Trends Plant Sci 9: 26-32.
38. Sharma A, Sahgal M, Johri BN. 2003. Microbial communication in the rhizosphere: Operation of quorum sensing. Curr Sci 85:1164-1172.
39. Velicer GJ. 2003. Social strife in the microbial world. Trends Microbiol 11:330-337.
40. Rice SA, Givskov M, Steinberg P, Kielleberg S. 1999. Bacterial signals and antagonists: the interaction between bacteria and higher organisms. J Mol Microbiol Biotechnol 1:23-31.
41. Sperandio V, Torres AG, Jarvis B, Nataro JP, Kaper JB. 2003. Bacteria-host communication: the language of hormones. Proc Natl Acad Sci USA 100: 8951-8956.

42. Margulis L. 1996. Archaeal-eubacterial mergers in the origin of Eukarya: Phylogenetic classification of life. Proceedings of the National Academy of Sciences of the USA 93: 1071-1076.
43. Margulis L. 1999. Die andere Evolution. Heidelberg: Spektrum Akademischer Verlag.
44. Margulis L. 2004. Serial endosymbiotic theory (SET) and composite individuality. Transition from bacterial to eukaryotic genomes. Microbiol Today 31:173-174.
45. Margulis L, Sagan D. 2002. Acquiring Genomes. A Theory of the Origin of Species. New York: Basic Books.
46. Odintsova MS, Yurina NP. 2005. Genomics and Evolution of Cellular Organelles. Russ J Gen 41:957-967.
47. Witzany, G. 2006. Serial Endosymbiotic Theory (SET): The biosemiotic update. Acta Biotheoretica 54:103-117
48. Batut J, Andersson GE, O'Callaghan D (2004) The evolution of chronic infection strategies in the *a*-proteobacteria. Nat Rev Microbiol 2:933-945.
49. Blech J. 2000. Leben auf dem Menschen. Die Geschichte unserer Besiedler. Hamburg: Rowohlt Taschenbuch Verlag.
50. Federle MJ, Bassler BL. 2003. Interspecies communication in bacteria. J Clin Investig 112:1291-1299.
51. Dunn AK, Handelsman J. 2002. Toward an understanding of microbial communities through analysis of communication networks. Antonie van Leeuwenhoeck 81:565-574.
52. McNab R, Lamont RJ. 2003. Microbial dinner-party conversations. The role of LuxS interspecies communication. J Med Microbiol 52:541-545.
53. Ben Jacob E, Levine H. 2006. Self-engineering capabilities of bacteria. J R Soc Interface 3:197-214.
54. Velicer GJ, Kroos L, Lenski RE. 2000. Developmental cheating in the social bacterium Myxococcus Xanthus. Nature 404:598-601.
55. Ben Jacob E, Becker I, Shapira Y, Ledvine H. 2004. Bacterial linguistic communication and social intelligence. Trends Microbiol 12:366-372.
56. Shapiro JA. 1998. Thinking about Bacterial populations as multicellular Organisms. Annu Rev Microbiol 52:81-104.
57. Velicer GJ, Yu YT. 2003. Evolution and novel cooperative swarming in the bacterium Myxococcus Xanthus. Nature 425:75-78.

58. Kreft JU. 2004. Biofilms promote altruism. Microbiology 150:2751-2760.
59. Taga ME, Bassler BL. 2003. Chemical communication among bacteria. Proc Natl Acad Sci USA 100:14549-14554.
60. Wang LH, He Y, Gao Y, Wu JE, Dong YH, He C, Wang SX, Weng LX, Xu JL, Tay L, Fang RX, Zhang LH. 2004. A bacterial cell-cell communication signal with cross-kingdom structural analogues. Mol Microbiol 51:903-912
61. Kaiser D, Welch R. 2004. Dynamics of fruiting body morphogenesis. J Bacteriol 186:919-927.
62. Kohlenbrander PE, Andersen RN, Blehert DS, Egland PG, Foster JS, Palmer RJ. 2002. Communication among Oral Bacteria. Microbiol Mol Biol Rev 66:486-505.
63. Kohlenbrander PE, Egland PG, Diaz PI, Palmer RJ 2005. Genome-genome interactions: bacterial communities in initial dental plaque. Trends Microbiol 13:11-15.
64. Rickard AH, Palmer RJ jr, Blehert DS, Campagna SR, Semmelhack MF, Egland PG, Bassler BL, Kolenbrander PE. 2006. Autoinducer 2: a concentration-dependent signal for mutualistic bacterial biofilm growth. Mol Microbiol 60:1446-1456.
65. Moore WEC, Moore LVH. 1994. The bacteria of periodontal diseases. Periodontology 5:2000.66-77.
66. Kroes I, Lepp PW, Relman DA. 1999. Bacterial diversity within the human subgingival crevice. Proc Natl Acad Sci USA 96:14547-14552.
67. Paster BJ, Boches SK, Galvin JL, Ericson RE, Lau CN, Levanos VA, Sahasrabudhe A, Dewhirst FE. 2001. Bacterial diversity in subgingival plaque. J Bacteriol. 183:3770-3783.
68. Imaizumi-Anraku H, Takeda N, Cherpentier M, Perry J, Miwa H, Umehara Y, Kouchi H, Murakami Y, Mulder L, Vickers K, Pike J, Downie A, Wang T, Sato S, Asamizu E, Tabata S, Yoshikawa M, Murooka Y, Wu GJ, Kawaguchi M, Kawasaki S, Parniske M, Hayashi M. 2005. Plastid proteins crucial for symbiotic fungal and bacterial entry into plant roots. Nature 433:527-531.
69. Xie G, Bonner CA, Song J, Keyhani NO, Jensen RA. 2004. Intergenomic displacement via lateral gene transfer of bacterial trp operons in an overall context of vertical genealogy. MBC Biol 2: 15.

70. Thomas CM, Nielsen KM. 2005. Mechanisms of, and barriers to, horizontal gene transfer between bacteria. Nat Rev Microbiol 3:711-721.
71. Bordenstein SR, Reznikoff WS. 2005. Mobile DNA in obligate intracellular bacteria. Nat Rev Microbiol 3:688-699.
72. Gogarten JP, Townsend JP. 2005. Horizontal gene transfer, genome innovation and evolution. Nat Rev Microbiol 3:679-687.
73. Jain R, Rivera MC, Lake JA. 1999. Horizontal gene transfer among genomes: The complexity hypothesis. Evolution 96: 3801-3806.
74. Jain R, Rivera MC, Moore JE, Lake JA. 2003. Horizontal Gene Transfer accelerates genome innovation and evolution. Mol Biol Evol 20: 1598-1602.
75. Brown JR. 2003. Ancient horizontal gene transfer. Nat Rev Gen 4: 122-132.
76. Berg OG, Kurland CG. 2002. Evolution of Microbial Genomes: Sequence Acquisition and Loss. Mol Biol Evol 19: 2265-2276.
77. Rivera MC, Lake JA. 2004. The ring of life provides evidence for a genome fusion origin of eukaryotes. Nature 431: 152-155.
78. Kunin V, Goldovsky L, Darzentas N, Ouzounis CA. 2005. The net of life: Reconstructing the microbial phylogenetic network. Gen Res 15: 954-959.
79. Tettelin et al. 2005. Genome analysis of multiple pathogenic isolates of Streptococcus agalactiae: implications for the microbial "pan-genome". PNAS 102: 13950-13955.
80. Graumann PL. 2004. Cytoskeletal elements in bacteria. Curr opin Microbiol 7: 565-571
81. Graumann PL, Defeu-Soufo HJ. 2004. An intracellular actin motor in bacteria? Bioessays 26:1209-1216.
82. Defeu-Soufo JH, Graumann PL .2004. Dynamic movement of actin-like proteins within bacterial cells, EMBO Rep 5:789-794.
83. Villarreal LP. 2005. Viruses and the evolution of life. Washington: American Society for Microbiology Press.
84. Forterre P. 2002. The origin of DNA genomes and DNA replication proteins. Current Opinion in Microbiology 5:525-532.
85. Forterre P. 2005. The two ages of the RNA world, and the transition to the DNA world: a story of viruses and cells. Biochimie 87:793-803.

86. Forterre P. 2006. The origin of viruses and their possible roles in major evolutionary transitions. Virus Res 117:5-16.
87. Villarreal LP, DeFilippis VR, 2000. A hypothesis for DNA viruses as the origin of eukaryotic replication proteins. Journal of Virology 74:7079-7084.
88. Villarreal LP. 2004. Can Viruses Make Us Humans? Proceedings of the American Philosophical Society 148:296-323.
89. Villarreal LP, Defilippis VR, Gottlieb KA. 2000. Acute and Persistent Viral Life Strategies and Their Relationship to Emerging Diseases. Virology 272:1-6.
90. Eigen M, Winkler R. 1975. Das Spiel – Naturgesetze steuern den Zufall. München: Pieper.
91. Witzany G. 1995. From the ‚Logic of the Molecular Syntax' to Molecular Pragmatism. Explanatory deficits in Manfred Eigen's concept of language and communication. Evol Cog 1:148-168
92. Takemura M. 2001. Poxviruses and the origin of the eukaryotic nucleus. J Mol Evol 52:419-425.
93. Bell JL. 2001. Viral Eukaryogenesis: Was the Ancestor of the Nucleus a complex DNA virus? J Mol Evol 53:251-256.
94. Bell PJL. 2006. Sex and the eukaryotioc cell cycle is consistent with a viral ancestry for the eukaryotic nucleus. Journal of Theoretical Biology doi: 10.1016/j.jtbi.2006.05.015
95. Nandhagopal N, Simpson AA, Gurnon JR, Yan X, Baker TS, Graves MV, Van Etten JL, Rossmann MG. 2002. The structure and evolution of the major capsid protein of a large, lipid-containing DNA virus. PNAS 99:14758-14763.
96. Rice G, Tang L, Stedman K, Roberto F, Spuhler J, Gillitzer E, Johnson JE, Douglas T, Young M. 2004. The structure of a thermophilic archaeal virus shows a double-stranded DNA viral capsid type that spans all domains of life. PNAS 101:7716-7720.
97. Khayat R, Tang L, Larson ET, Lawrence CM, Young M, Johnson JE. 2005. Structure of an archaeal virus capsid protein reveals a common ancestry to eukaryotic and bacterial viruses. PNAS 102:18944-18949.

---

* co-author: Pierre Madl; Institute of Physics & Biophysics at the University of Salzburg

5

Natural Genome-Editing Competences of Viruses*

Abstract: It is becoming increasingly evident that the driving forces of evolutionary novelty are not randomly derived chance mutations of the genetic text, but a precise genome editing by omnipresent viral agents. These competences integrate the whole toolbox of natural genetic engineering, replication, transcription, translation, genomic imprinting, genomic creativity, enzymatic inventions and all types of genetic repair patterns. Even the non-coding, repetitive DNA sequences which were interpreted as being ancient remnants of former evolutionary stages are now recognized as being of viral descent and crucial for higher-order regulatory and constitutional functions of protein structural vocabulary.

In this article I argue that non randomly derived natural genome editing can be envisioned as (a) combinatorial (syntactic), (b) context-specific (pragmatic) and (c) content-sensitive (semantic) competences of viral agents. These (three-leveled) semiotic competences could explain the emergence of complex new phenotypes in single evolutionary events.

After short descriptions of the non coding regulatory networks, major viral life strategies and pre-cellular viral life three of the major steps in evolution serve as examples: There is growing evidence that natural genome-editing competences of viruses are essential (1.) for the evolution of the eu-

* Acta Biotheoretica (2007). doi: 10.1007/s10441-006-9000-7.

karyotic nucleus (2.) the adaptive immune system and (3.) the placental mammals.

Keywords: non-coding DNA, repetitive elements, genome editing, persistent viral life-style.

1. Introduction

Significant evolutionary modifications can occur by changing the repetitive elements that format the genome architecture rather than, as has been widely assumed so far, by altering protein-coding sequences. This implies that changes in the regulatory framework may lead to new complex phenotypes without even requiring a modification of the coding sequences.

For decades, non-coding regions of the genome have been ignored or declared as "junk"-DNA. Over the past decade, however, scientists have realized that these regions incorporate decisive higher-order regulatory functions. Within the human genome sequences are only 3 % coding for proteins and 97% non-coding ones. But these 3% of the protein-coding sequences match with those of the mouse by up to 99%. Highly developed genomes among mammals contain a nearly identical protein-coding vocabulary.

While bacterial genomes almost lack such repetitive elements, there is a tenfold increase of such segments in the genome of the eukaryotic yeast cells; in the case of the fruit fly *Drosophila*, the repetitive elements already amount to 10%, while in humans they reach 97%.[1] These formatting, repetitive elements are a prerequisite for the decisive DNA-editing processes such as expression, replication, repair and recombination.

2. Non-coding regulatory networks and genome editing

Clearly, mobile sequences such as transposons and retroposons[2] and non-coding repetitive elements such as LTRs, SINEs and LINEs (long terminal repeats, short interspersed elements, long interspersed elements) enable far-reaching DNA rearrangement and reorganization.[3-6] Together, they play a decisive role in the evolution of new genomic structures.[7-10] Depending on the organism's state of development, the varying chromatin markers are, thus capable – through different methylation patterns, histone modifications and alternative splicing – of creating a set of "multiple protein meanings"[11] from one and the same genetic data-set.[12-19] This even characterizes the rise of epigenetics, i.e. the view that phenotypic variations, which are heritable, need not be connected with genetic alterations.[20-24] The question arises as to how and why the evolution of higher genetic complexity is connected to non-coding DNA, formerly termed "junk"-DNA?

Although we have known for several decades that the unbelievable diversity of enzyme proteins is a practical tool for DNA editing processes, it was unclear by which rules or higher-order regulations they are governed.[25-27] It subsequently became clear that higher-order regulations such as co-suppression, suppression of transposition, position effect variegation, several start- and stop-signals, RNA interference, imprinting, chromosomal methylation, transvection and transcriptional and post-transcriptional gene silencing are processed by micro-RNAs.[28-31] In 1992 Watson suspected these higher-order regulatory functions to be hidden in chromosomal structures.[32] Interestingly, researchers analyzing the linguistic features of non-coding DNA sequences draw the right conclusions.[33,34] They found coherence between coding DNA with similar structural features as artificial scientific languages and non-coding DNA with similar structural features as everyday language.

The genome, with both its coding and non-coding sections, serves as a genetic, epigenetic and computational cell storage medium that involves different kinds of storage techniques. These include short-time storage (dynamic rearrangements, remodifications, rapidly induced alterations for expression and repair), intermediate storage (epigenetic use) including multiple protein-meanings within the same chromosomal structures by different expression modes through different methylation patterns, and long-time storage, i.e. stable inheritable datasets for many generations.[5] This undoubtedly means that the DNA storage medium in its possible functions depends on different means, i.e. interpretations, of the organism inputs on this medium rather than on the inherent information. This enables changing inheritable priorities, whether computational, intermediate, or long-time storage data – or combinatorial patterns among them.[35,20,36]

The repetitive sequences are highly species-specific and more suitable for the determination of species than the coding sequences.[37] Each taxon organizes and formats its genome architecture differently, i.e. regulation of expression, transcription, replication, and translation are species-specific. Within each taxon, these processes, along with the associated gene architecture, must run in a highly coordinated manner so that they do not disturb each other. Only a co-ordination of the individual steps guarantees that these different actions are performed and maintained successfully: DNA sequences must be read at the right site and at the right time. The precise spatio-temporal co-ordinations are essential in order to sustain vital processes; nonetheless, such co-ordinations can also fail, either by sequence-damage (i.e., mutations) or by organism-induced translational, transductional, repairing or other rearrangement disturbances.

Experience has shown that (1) excision, (2) insertion and (3) combination of the genomic codes are the keys in DNA editing and the basis of evolutionary processes. The contemporary term would be "natural genetic engineering".[8] Wittgenstein noted that "to understand a sentence means to understand a language. To understand a language means to be master of a technique".[38] Nonetheless, there is a clear difference between natural and artificial genetic engineering. The former provides ontologically genuine products that are evident in all living beings and life processes, whereas the latter attempts to achieve modifications and improvements by copying the natural genome-editing competences.

3. Major viral life-stretegies

New research has shown that these non-coding repetitive sequences originated primarily from retroviral RNA.[39,37] After persistent non-lytic viral infection, important protein coding regions become incurred within the replicating region of the host organism, whereas repetitive sequences are integrated into the non-coding sections.[40,41] In this way, thousands of endogenous retroviral sequences have been integrated into the human genome, with 22 independent retroviral families currently having been identified.[39,37,36] The quantity of remaining former viral-gene embedding repetitive elements (embracing an enormous genetic diversity) originally accompanied the protein-coding sequences as control- and/or identification segments. Most endogenous retroviruses have been degraded into formerly connected domains, but they can still be recognized by their three genes *gag, pol* and *env.*[42,36] The *gag* gen encodes structural proteins, *pol* encodes enzymes such as reverse transcriptase and integrase functions and *env* encodes envelope proteins.

At this stage we know roughly 3600 different kinds of viruses and approx. 30,000 viral subtypes.[37:6] The most numerous variety is found among single-stranded RNA viruses, followed by double-stranded DNA viruses, double-stranded RNA viruses and single-stranded DNA viruses.[37:5] This huge variety of viruses and viral components is even more astonishing when considering the fact that 1ml of sea water contains about 1 million bacteria and that the number of viral components in the same amount of water is approx. 10 times higher.[37:55] However, their ecological niches are not just limited to purely aquatic or terrestrial environments, but also include species-specific tissues that differ significantly from each other.[37:56] As phylogenetic analysis shows, nearly all organisms of all kingdoms have been infected or highly colonized by viruses since the beginning of life.

Viruses can parasitize almost any replication system – even prebiotic ones – and probably emerged well before the appearance of cellular life forms. Viruses store information that not only pertains to (a) replication proteins but also to (b) morphology and (c) phenotypic diversity. Based on this and the results of phylogenetic analyses and comparative genomics, it is possible to establish viral lines of ancestral origin. These lines of origin can also be non-linear because different parts of viruses contain different evolutionary histories.[37:11]

Since viruses with RNA genomes are the only living beings that use RNA as a storage medium, they are considered to be witnesses of an earlier RNA-world, of a time when DNA did not exist yet.[43-47] Negatively stranded RNA viruses have genome structures and replication patterns that are dissimilar to all known cell types. As demonstrated by phylogenetic analyses, cellular replicases are related to each other; however, there is no similarity between RNA-viral replicases and those of any known cell types. This proves the existence of negatively stranded viral RNA-replicases even before cellular life came into being.[37:11] DNA viruses, too, do not give any reference to a cellular origin. Phylogenetic analyses point to an older time scale, as DNA-repairing proteins of DNA viruses do not have any counterparts in cellular biota.

For a long time, viruses were not considered a part of the animated world because they were believed to be exclusively parasitic in nature. They were interpreted as causing an acute infection of the host organisms, using the host's cellular machinery to reproduce, and performing their lytic nature only in order to infect other cells. Although this narrative remains valid, it merely represents a special case of viruses that were unable to reach persistent status by a sessile life-style.[39:305] Most viruses, however, are stable, persistent living beings that do not colonize a host organism for simple selfish purposes. Rather, they display varying survival strategies which can differ according to the host they inhabit:[37:7]

- In bacteria we mainly find double-stranded DNA viruses. Similar viral types are also found in algae, but they are absent in fungi and plants.
- Fungal hosts house mainly double-stranded RNA viruses, whereas
- Plants contain predominantly single-stranded RNA viruses.
- Mammals, on the other hand, are colonized mainly by endogenous retroviruses (ERVs).

Viral persistence in host organisms is crucial because they reliably protect the host against similar parasites. Interestingly, competing viruses are in most cases of the same or related species, whereas unrelated viruses do not compete but interact symbiotically.[48]

As for endogenous retrovirus, they confer the host a distinct and unique genetic identity, non-existent before the colonization. This persistence can either be endonucleic or cytoplasmatic. Persistent viruses are capable of continuous or episodic reproduction in the hosts.[37:5] The main direction is not to infect cells for reproduction with lytic consequences and often epidemic viral diseases, but a sessile life-style without harming the host. This includes different meanings of the term fitness. Thus, lytic life-styles are fit if the reproduction rate reaches epidemic dimensions, whereas the persistent life-style is fit if it does not harm the host and wards off competitive viral infections.

For far too long, viruses have only been considered as poisonous and highly infectious parasites. The fact that viruses are silent companions of virtually all organisms and that they play a decisive role in the evolution of the host has been largely ignored.

Gene functions of cellular biota, which undoubtedly are associated with the persistent life strategy of viruses, include:[37:21]

- Immunity (restriction and modification modules, toxic and antitoxic modules);
- Silencing functions/micro-RNAs, (methylation, suppression);
- Recognition functions (replicate expression, receptors, expression factors);
- Immune regulation (signal mediating, heredity, adaptation).

However, the host must be able to inhibit the replication of a persistent virus; i.e., it must prevent it from turning aggressive and exercising its lytic nature. Especially if the persistence does not reach a stable status, the move-countermove interaction between nucleic acid invader and a host genome may be an on-going process which leads to a special kind of immunity, a defensive RNA-silencing against competing viruses.[49,50] This RNAi interactional pattern is documented to be a very old type of innate immune system.

4. Examples of diverse strategic patterns of viruses

4.1 Virus escape-strategy

Mitochondria and chloroplast precursors take refuge in prokaryotic hosts where they are protected from cyanophagous colonization. Although the host is equipped with restriction and modification competence that can ward off viral colonization, endosymbiotic mitochondria and chloroplasts lack such competences. Such competences might have existed in the past, but they have been relocated to the hosting organism, thereby establishing a type of immune system against cyanophagous infestation, which is omnipresent in oceans.[37:127]

4.2 Wall-off-Strategy

This is closely associated with calcification (encapsulation of the parasites) and provides some mechanical as well as chemical protection. It was initially based on an RNAi defence system found even among simple marine species such as early protostomes and among their relatives such as crustaceans and insects. As in bivalves, this phenomenon is also found among insects: a chitinous or a mineralized barrier is formed around the parasitic intruder. This is a very archaic viral strategy that can still be observed today as a reaction to parasitic infections in mammals and in the form of a chronic inflammatory response in vertebrates.[37:204]

4.3 Addiction module

Addiction modules include antagonistic components acquired in order to survive inside a host without damaging it.[37:302] Acquisition of addiction modules can be reconstructed as a massive viral colonization of the host. Addiction modules consist of a set of genes or functions that are harmless to the host and that include both toxin and antitoxin. In most cases the evolutionary aspect for the host organism is an acquired and hereditary immunity function, e.g. the restriction/modification phenotype. These modules probably paved the way for the evolution of new species, so persistent colonization led to permanent integration without being threatened by any other virus;[37:145] e.g. *Dictyostelium,*[37:178,181] sharks, bryophytes[37:238] and ferns;[37:228] this process takes place even today. However, the present diversity of these innovative phenotypes is in turn immensely colonized by viruses and, interestingly, by groups of novel viruses that were non-existent before.

4.4 Multiplicity reactivation

Some of the oldest viruses, the phycodnaviruses and cyanophages, are able to repair UV-light induced damages of their genomes and can reproduce even if the host organisms are exposed to lethal doses of radiation. This ability to replicate is only possible due to active participation of virus-specific repair enzymes, which can re-establish the cell's competence to process macromolecules.

This capability of expressing subsets of genes is retained even if each of the viruses' genome experiences lethal radiation damage. Once functional complementation of expressed genes and other damaged viruses is established, the complementary sets are capable of initiating viral replication: the necessary combinations of defective viral genomes are utilized in order to assemble intact viruses. Therefore, viruses are the only living beings capable of meaningfully recombining text fragments of a damaged genome into a fully functional viral genome capable of self-replication.[37:128]

4.5 Sexual isolation

Once a species is massively infested by an acute and lytic pandemic virus, some of the surviving hosts manage to permanently include this viral genome into their own genome, thereby obtaining permanent immunity against such infections. As soon as they engage in intraspecific reproduction, they lethally infect those siblings that lack this additional immunity. This sexually isolates the immunized from the non-infected members. As an immediate result, the genetically altered siblings experience an interruption of their line of ancestry by acquiring a complex immunological phenotype. This enables the immunized species to establish a novel species incapable of mating with the previously related members without killing them.[39:315;37:361]

5. Pre-cellular viral life

RNA genomes probably existed before the appearance of DNA genomes.[37:30] Hence, DNA can be considered to be a modification of RNA, yielding ribonocleotide reductase, followed by only two thymidylate synthases.

Initially, there were only two components, RNA molecules and RNA proteins, which constituted a RNA cell that contained only few, individual DNA components. In turn, they were colonized by parasitic RNA that facilitated the scission into two separate lineages – DNA viruses and RNA viruses – whereby a DNA virus was capable of infecting an RNA virus. This model would at least account for the existence of viral encoded DNA transaction proteins, to which no cellular counterpart exists. In addition, this model would explain the existence of two dissimilar DNA-replication systems.

On the one hand, the absence of repair enzymes within the RNA world gave it a more flexible and creative character. On the other hand, this inability is a drawback when a reliable storing capacity is required. Evidently, the presence of such addiction modules within the highly competitive viral RNA-DNA world must have an advantage. This is particularly true when a DNA virus infects an RNA virus, forcing the latter to establish a bi-layered cell-membrane and to encapsulate the genome by a porous nuclear envelope. Doing so still enables replication, repair and recombination.[37]

Even though the DNA cell gained distinct selective advantages, the RNA parasites still had an astonishingly powerful genomic creativity;[36] this imparted distinct survival advantages if environmental conditions changed considerably. Today we know that the DNA world, on its own, would not have brought forth such an incredible diversity by natural selection[51]– let alone established the necessary genetic precondition to create such a high degree of complexity. Overall, the genomic innovation of the RNA world complements that of the more conservative and stable DNA world.

Besides these processes, life also resulted from two alternative innovations: the creation of the archaeal domain and that of the eubacterial domain. Again, this would explain why archaeal lipids are so much different from eukaryotic/bacterial lipids. These lipids contradict eubacterial stereochemistry and differ in the backbone architecture of the hydrocarbon chain.

6. The origin of the eukaryotic nucleus

The origin of the eukaryotic nucleus serves as a first example for the important role of natural genome-editing competences of viruses. Since the introduction of the Serial Endosymbiotic Theory (SET), it is generally accepted that the eukaryotic cell did not result from random mutations, but from the coordinated union of free-living prokaryotes.[52-57] Mitochondria and other organelles clearly descended from these micro-organisms [58,59] and the assumption was that the eukaryotic nucleus is also probably of archaeal or bacterial descent.

New evidence supports the idea that eukaryotic nuclei originated before the symbiogenetic integration with mitochondria and chloroplasts.[37:102] In fact, the nucleus has basic properties that are otherwise absent in eukaryotic cells.[60,61]

Prokaryotes do have circular chromosomes with uniform, standardized origins of replication. Their chromosomes are only loosely attached to chromatin proteins and have different control regions that coordinate and terminate DNA replication. All eukaryotic proteins involved in DNA replication differ from those found in prokaryotes. Hence, nuclear properties of eukaryotes are completely different from those of prokaryotes.[39:306] These differences include:

- use of linear chromosomes, with repetitive termination points and several origins for replication,

- transcription and translation separated by multiple membranes,
- existence of complex nuclear pore structures that actively mediated RNA translocation,
- a tubulin-system that enables separation of duplicated chromosomes.

All these properties represent complex phenotypes that require complex co-ordination of numerous protein functions. None of these functions are present in prokaryotes even though they are considered as predecessors of the eukaryotic nucleus.

The eukaryotic nucleus contains three kinds of DNA-dependent RNA polymerases that differ significantly from RNA polymerases of prokaryotes.[37:106] Even the three kinds of splicing group I-introns (DNA transposase, reverse transcriptase and micro-RNAs) are largely inexistent in prokaryotes, although they are present in viruses of prokaryotes. Bacteria never had introns in any of their coding genes. In addition, no single prokaryotic process is known to account for the tasks of membrane disintegration and restoration as observed in eukaryotes. Eukaryotic tubulin consisting of a highly conservative mitotic spindle is absent in all prokaryotic lines of descent, even though it is one of the most important components in eukaryotes. Prokaryotes lack a tubulin system. The eukaryotic pore-structure of the nuclear envelope likewise has no counterpart in the prokaryotic world.[37:119]

Viral proteins on the other hand bind both tubulin as well as actin, thereby triggering polymerisation and mobility functions. Viral genes are directly involved in transposition tasks. All viruses mark their genome, their RNA and their proteins with virus-specific methylases, for example via enzymatic reactions known as base methylation.[37:122]

Viruses were long thought to be parasites capable of "stealing" host-specific properties. In the meantime, viruses are known to have created new genes as a result of their evolutionary line of descent. In the baculoviruses, for example, GenBank database investigations reveal that 80% of their genes are unique to this group and found nowhere else.[37:274] Gene losses have been documented in baculoviruses, but the 12 losses documented therein are opposed by a staggering acquisition of 255 new genes.

Between 1950 and 1980, scientists recognized that the T4 phage-polymerase proteins are much more similar to the eukaryotic DNA polymerase protein than to any prokaryotic polymerase. We now know that the eukaryotic DNA polymerase and the T4 DNA polymerase do have common origins. Indeed, the T4 represents a huge family of viruses that is capable of infecting both Bacteria and Archaea.[39:307] Hence, it is not surprising that T4-DNA polymerases are found in all three domains of life: Archaea, Bacteria and Eukarya.

Algae were among the first higher eukaryotic organisms that had to deal with viruses. Thus, viruses that infected microalgae must have had a large adaptive potential that accompanied the evolutionary pathway and must have included the characteristic of a protonucleus. The phycodnavirus is a case in point: this calls for examining the entire GenBank database for sequences that may be similar to the DNA polymerase of this particular virus.[39:308] Such sequences must include:

- replication polymerases of all higher eukaryotes as well as of all larger eukaryotic DNA-viruses,
- primer polymerases of eukaryotes,
- repair polymerases of both Archaea and Bacteria.

The DNA polymerase of the CSV1 virus is present at the origin of all eukaryotic replication DNA polymerases and is no doubt a precursor of all polymerases involved in replication of the eukaryotic genome. So far no other viral or prokaryotic DNA polymerase that shares these features is known.[39:310]

The membrane-bound separation of transcription and translation is a characteristics of the poxviruses, more specifically of the Vaccina and other DNA viruses. Moreover, these viruses have a very simple pore-structure that has actively been incorporated from the membrane-bound RNA into the host cytoplasm. A similar situation can be documented with the small chromatin proteins and the linear chromosomes along with their repetitive telomere tails that are so characteristic among various cytoplasmid DNA-viruses, TTV1 and phycodnaviruses. Even the highly complex function of tubulin, as an important coordinating element during chromosomal separation of duplicated strands, is present in DNA-viruses with exactly the same set of functions.[39:311]

It became increasingly obvious that all properties of the eukaryotic nucleus are derived from a large, stable and persistent DNA virus with linear chromosomes. The current interpretation is that the precursor of the eukaryotic nucleus was indeed a huge membrane-covered DNA virus that persistently colonized a prokaryotic host.[39:311] Therefore, the hosting cell must have lost its cell wall, with the virus incorporating the prokaryotic genes into its pre-nuclear genome, particularly in cases of encoding for metabolism and translation. This virus was most likely non-lytic because it coordinated both its own replication and its transcription genes, it had a double-layered membrane and a tubulin system in order to wrap chromosomes. Its persistence and its reactivation would imply that (a) the process of cell division (nuclear envelope dispersion and reformation), (b) mitotic duplication (doubling of the chromosomes and allocating them to the progeny cells) and (c) the viral DNA corresponds to the sexual reproductive cycle of the host organism.

Interestingly, all these properties can be found in prokaryotic viruses such as cyanophage, archaeal phage, mycobacterial phage and eubacterial phage.[37:112]

7. The origin of the adaptive immune system

The second example for the important role of viral genome-editing competences is the origin of the adaptive immune system. In the case of persistent viruses, the host must actively be able to take control over viral replication. This requires the host to have some kind of defence/immune system.[37:209] On the other hand, the persistent virus must also be able to wall off similar viruses trying to take advantage of the host. Lower organisms achieve this with so-called addiction modules – sets of genes that impair the host when absent but are of advantage when present in the host.[39:302] Usually, the negative aspects of such gene sets involve toxic substances, while the positive involve the 'antidote'. At any rate, acute and lytic action might be encountered among closely related species, provided that they are persistent to a specific host. Only in such cases is the hosting organism protected. Such virus-based immune modules are complex genetic systems that include interrelated functional sets that can positively or negatively affect the host; certain environmental conditions or colonization by other viruses can alter the current persistent state by becoming acute, lytic and thereby damaging or even killing those that are not yet infected.

Prokaryotes of the archaeal and bacterial domain are the most adaptive individuals on earth. This raises an obvious question: how do these organisms protect themselves against viral agents? The answer involves utilizing restriction/modification enzymes: Restriction enzymes degrade unmodified DNA with modification enzymes acting only during DNA replication. This simple and innate immune system likewise originated from viruses. Prokaryotes colonized by such viruses possess specific immune response capabilities. The host organism acquires new complex immunologic properties once the viral colonization stabilizes.[39:303]

This, however, does not involve horizontal gene transfer (HTG), but rather large-scale viral inventions of new genes. Some find their way through host lineages and become permanently established within the host. Others have to assess the molecular texture of the bacterial genome as well as its evolutionary potential.[39:305]

A major change in evolution occurred during the transition from urochordates (sea squirts/tunicates) to bony fish. Many common tissue types are still evident, although the bony structure required a huge increase in the genome. This was predominantly achieved by retroposons, which originated from endogenous retroviruses. Urochordates, on the other hand, possess only an innate immune system.

7.1 The acquisition of a complex new phenotype

Bony fish are the first in this line of descent that possess both an innate and an adaptive immune system.[37:203] The conclusion is that the entire complexity of an adaptive immune system was acquired at once, probably at the very beginning of the vertebrate line of descent. At the same time, evolution simultaneously yielded jawbones, the vertebral spine, and the skull. Concomitantly, new viral families emerged that had never inhabited any of the previous life forms: four different kinds of RNA-virus families, non-defective, autonomous and abundant retroviruses.[37:204] The adaptive immune system therefore represents an interlinked network of proteins that tag inflammation and other acute processes (cytokine and chemokine), and their receptors and signal transmission systems, that stimulate the humeral and cellular antigen-specific immune response pattern. This must have been acquired in a single event because it is monophyletic.[37:205]

While urochordates do not possess any – except for one – of these features, bony fish possess them all. Sharks are among the earliest vertebrates that featured an adaptive immune system.[37:206] Interestingly, hardly any viruses are found in this group (except for a herpes type strain). They have a very rudimentary, adaptive immune system. Sharks are among the first organisms that include both oviparous as well as viviparous species.[37:207]

The reverse side of an adaptive immune system has a drawback: the non-related haploid chromosomal set is no longer rejected. From an immuno-related perspective it is not recognized as being "alien" – very much in analogy to the more recent development of placental mammals. Nonetheless, a large quantity of cytokines are transcribed within the placenta, i.e. the immune system is highly active, as is the complementary function of the addiction module. Because the mechanisms of self-nonself-identification are extremely vigilant, they protect the forming progeny from external parasitic attacks.

The acquisition of an adaptive immune system is both a punctual and a variable evolutionary event of the animal kingdom. It enabled the expression of highly complex phenotypes. These phenotypes consist of a self-forming and a dynamically adapting genetic system that recognizes "non-self" elements and can thereby promptly attack and simultaneously prevent fatal auto-aggression. In the context of a manifold self-identification system, this acquisitional gene-set strategy was developed in order to detect novel non-self agents. Once the system recognizes the presence of "non-self" agents, it responds by developing a new molecular process that involves the generation of genetic diversity and clonal growth of specific cells capable of detecting such non-self agents. This kind of genetic recombination on a genetic level is found only from this point onwards, not in any predecessors.[37:209]

The overall result of such processes is cells that produce new classes of molecules that can (a) bind and suppress non-self agents or (b) enable amoeboid cytotoxic cells to find and neutralize agent-containing cells. Most of these characteristic properties of an adaptive immune system were lacking prior to vertebrate evolution.

Today, we know that tunicates, the precursors of urochordates, possess a polymorphous MHC-like system which is associated with cell-induced, non-adaptive, amoeboid hemolymphic lethal actions.[37:209] This tunicate system, however, lacks a molecular similarity to the vertebrate MHC system. In order to develop an adaptive immune system so common to vertebrates, the tunicate-like system had to acquire an adaptive component that allowed the non-self elements to be detected as alien. Accordingly, an adaptive immune system destroys non-self elements while protecting those that are part of the system.

These related properties are characteristic for addiction modules. Indeed, the adaptive immune system is an elaborated addiction module. Hosts which incorporated it acquired a system with destructive capabilities. This ability to kill is comparable to a potent toxin that lyses any cell exposed to it. Through this self-recognition process the host must be equally able to prevent autolysis, comparable to an antitoxin.[37:210]

Similar to other addiction modules, the (a) lethal, toxic ability of the adaptive immune system is stable and long-lasting, while (b) the ability to express antitoxic characteristics (self-recognition) is only a temporary feature acquired during the development stage of the immune cell. Thus, the adaptive immune system reveals two aspects of an addiction module (toxic and antitoxic) and a varying stability of the toxin with regard to the protecting antitoxin.[37:210]

Many of these individual elements of an adaptive immune system evolved individually early on in evolution. They definitely existed previously in different viral types rather than in cellular organisms.

What is the most important basic gene-function of an adaptive immune system that favors diversity (necessary for recognizing non-self elements), and where did these genes evolve from? The proteins resulting from recombinase activating genes (RAGs) are responsible for DNA adaptations and new combinations.[37:211] They are the driving agents establishing genetic diversity, thereby producing a manifold display of varying surface receptors. From this perspective, RAGs are the crucial starter proteins for the selection of adaptive immunity. Phylogenetic analyses of these RAGs suggest that precursors existed neither in eukaryotic nor prokaryotic genomes. Instead, RAGs are probably closely related to integrase genes of retroviral origin.

7.2 Where can the ancestral origin of an adaptive immune system be found?

So far, no specific viral family that can be considered as a precursor of an adaptive immune system has been isolated. No single virus family possesses all the necessary and required properties. The basic ability to acquire RAG functions and their controlling capability is also accompanied by the property of endogenous retroviral colonization of the host genome.[37:212] The above properties point to the acquisition of adaptive immunity rather than a complex punctual genetic event. This acquisition event however, was not the product of an individual genetic parasite but rather the joining of the stable colonization through entire sets of supplementary and defective viral agents.[37:212] These agents must have covered the immediate precursors of cartilaginous fish with their extremely complex addiction module. This enabled both a stable colonization and successfully excluded competing parasites. The host-predecessor was most certainly equipped with a recognition system quite similar to MHC – a system of cytotoxic cells that was colonized and regulated by a complementary set of newly colonizing parasitic agents. This eventually resulted in the creation and evolution of the adaptive immune system.[37:213]

8. The evolutionary innovation of placental mammals through endogenous retroviruses

The third example for the important role of genome-editing competences of viruses is the origin of placental mammals. The close ties between the human genome and the colonization with repetitive sequences of retroviral origin (LTRs, SINEs, LINEs) become obvious when considering the Y-chromosome.[39:314] It contains only 20-30 coding sequences, approximately 225 genes,[37:338] whereby the majority are non-coding sequences with higher-order regulatory functions. The remainder of endogenous retroviruses is mainly found on the Y-chromosome. Interestingly, the genome-editing competence of Alu elements is evident in that they can change their own gene expression by modifying their own genomic meythylation status.[62,63]

The most active period of endogenous retroviral transcription occurs during the formation of placental tissue, during growth periods, and when trophoblasts join together.[39:314] Trophoblasts encapsulate the egg, help the egg nest properly, trigger processes that ensure nutrition, and prevent reactive responses by the mother's own immune system. The egg is protected by trophoblasts against an immuno-reactive response of the mother. These characteristics are unknown to monotreme mammals and marsupials. The acquisition of such abilities must have been a remarkable event.[39:314]

In turn, the trophectoderm is a very complex tissue that is, surprisingly, not of maternal origin, but a derivative of the fertilized egg – it even develops before the egg becomes implanted into the uteral lining. Experiments that suppressed expression of endogenous retroviruses inhibited implantation into the uteral lining. This implies that implantation of the embryo requires transcription of retroviral syncytin-coding genes. In humans, the HERV W *env* gene codes for syncytin,[64] a molecule used by the host to join trophoblast cells with the tissue that eventually nourishes the embryo.[39:315]

Although these processes have already been known for over 30 years, the purpose of this reaction was unclear: it did not make sense that the evolutionary innovation of placental mammals was tied to the acquisition of a complex set of endogenous retroviruses. Since the trophectoderm is protected by the maternal immune system, it enables further growth into the placenta, thereby modifying blood flow and nutrient supply between mother and embryo. The trophectoderm is associated with extremely high expression rates of ERV genes that result in RNAs as well as in other gene products and retroviral corpuscles. ERVs are highly host-specific and are closely associated with LINEs and SINEs of placental species. Their expression is not suppressed in the trophectoderm.

Once the sex of the totipotent embryo is determined, the high ERV expression rates are stopped and DNA methylation starts functioning.[37:325] Interestingly, ERV competences protect the embryo from the maternal immune system until the embryo's sex is determined. Only then does DNA methylation mediate between growth and development.

There are, however, clear references for evolutive and physiologically relevant qualities. For example, the expression of HERV-3 is boosted because it involves many fetal tissues in humans such as the adrenal cortex, kidney tubules, tongue, heart, liver and central nervous system as well as sebaceous glands of normal skin.[36] Thus, important tissues are formed during the fetal stage and are mediated via the presence of human endogenous retroviruses that were expressed during early mitotic divisions. They safeguard not only the formation of the placenta but also of the most important tissues of the fetus. If these retroviral components were removed from our genomes, we would be already extinct.

9. Summary

Phylogenetic analyses as well as GenBank database comparisons show that most natural genome-editing competences are not of cellular origin but represent original skills of viruses.

Far from being only stochastic results of randomly derived chance combinations as proposed earlier,[65-70] genome-editing competences are built on precise combinatorial rules (molecular syntax), highly sensitive to interactional contexts (pragmatic rules) that determine differences in the semantic content, e.g. multiple protein meanings generated according to higher-order regulatory functions.[26,71] genome editing of viruses depends primarily on their semiotic competences, i.e. their capability to generate and constitute genetic content as a language-like text according to syntactic, pragmatic and semantic rules in coherence with the laws of physics and chemistry. Through these competences the genetic code is used as a kind of innovation-sharing protocol.[72]

Viruses have two completely different life strategies, which are clearly reflected in their genomes. Accordingly, acute viruses exhibit lytic action and can induce disease and even death, whereas the life strategy of persistent viruses implies compatible interactions with the host. The latter are either integrated into the hosting genome or the cell plasma, and act non-destructively during most life stages of the host. The persistent life-style allows the virus to transmit complex viral phenotypes to the hosting organism. Doing so enables the host to broaden evolutive potentials that can lead to the formation of new species.

The natural genome-editing competences of viruses are most complex in bacteria, in which the complete nucleotide-word-order is largely determined, combined, and recombined by viruses. Hence, the main genomic novelties are found in the prokaryotic domain from where they originally evolved into the higher life forms. Probably all basic enzymatic variations originated therein (Villarreal 2006, personal communication). Massive viral colonization occurred from the very beginning of life, starting with the evolution of Bacteria and Archaea, later on of Protoctista and multicellular Eukaryota. The formation of all kingdoms, their families, genera, and species relies on the effects of viral colonization and results in diversified lineages and ultimately in the evolution of new species.

Increasing complexity and diversity involved genetic innovations, new combinatorial patterns of genetic content, non-coding regulatory networks and modifications of the genomic architecture. Interestingly, increasing complexity correlates with the expansion of non-coding DNA.[73] Combinatorial and rearranging processes in evolutionary and developmental processes occur non-randomly. They need successful and coherent adherence to the rules of molecular syntax.[74,25] Genetic content within the genomic matrix depends on the situational contexts in which living organisms are involved in vivo (e.g. growth, mating, virulence, stress, etc.). This content can therefore produce multiple protein-meanings, i.e. different semantic contents of the same genetic data-set. This is the prerequisite for epigenetically induced evolutionary and developmental processes. These rule-following processes may fail, with fatal consequences for the organism.[57] If evolutionary processes are intertwined between different species, then complexity is even more evident.[37,75] This indicates an important role of symbiogenetic processes in enhancing genetic, genomic and phenotypic complexity and diversity.

From this perspective, evolutionary history emerges entirely differently than previously thought. This history is not an aggregation of chance mutations of the genetic text and its associated selection. Rather, it is a permanent and competent processing of genetic sequences to acquire previously unknown abilities and to ward off competing parasites via genomic innovation. This explains why scientists had difficulties understanding evolutive patterns before comparative genomics was established: the thesis of natural genome-editing competences of viruses would merely have been another curious hypothesis. Contemporary sequence analyses and intercomparisons underline the crucial evolutionary functions of viruses and their genome-editing competences for all life forms.

References

1. Villarreal LP. 2004. Can Viruses Make Us Humans? Proc Am Phil Soc 148: 296-323.
2. Volff JN. 2006. Turning junk into gold: domestication of transposable elements and the creation of new genes in eukaryotes. BioEssays 28: 913-922.
3. Shapiro JA. 2002. Genome Organization and Reorganization in Evolution: Formatting for Computation and Function. In: L.v. Speybroeck, G.V.d.Vijver, Waele, DD. (eds). From Epigenesis to Epigenetics. The Genome in Context. Ann NY Acad Sci 981: 111-134.
4. Sternberg R. 2002. On the Roles of Repetitive DNA Elements in the Context of a Unified Genomic-Epigenetic System. Ann NY Acad Sci 981: 154-188.
5. Shapiro JA, Sternberg R. 2005. Why repetitive DNA is essential to genome function. Biol Rev 80: 1-24.
6. Deepak G, Majumder PP, Rao CB, Brahmachari SK, Mukerji M. 2003. Nonrandom distribution of alu elements in genes of various functional categories: insight from analysis of human chromosomes 21 and 22. Mol Biol Evol 20: 1420-1424.
7. Shabalina SA, Spiridonov NA. 2004. The mammalian transcriptome and the function of non-coding DNA sequences. Gen Biol 5: 105e.
8. Shapiro JA. 2004. A 21st century view of evolution: genome system architecture, repetitive DNA, and natural genetic engineering. Gene 345: 91-100.
9. Sternberg R, Shapiro JA. 2005. How repeated retroelements format genome function. Cyt Gen Res 110: 108-116.
10. Pollard KS, Salama SR, Lambert N, Lambot MA, Coppens S, Pedersen JS, Katzman S, King B, Onodera C, Siepel A, Kern AD, Dehay C, Igel H, Ares Jr M, Vanderhaegen P, Haussler D. 2006. An RNA gene expressed during cortical development evolved rapidly in humans. Nature, doi:10.1038/nature05113.
11. Ast G. 2005. The Alternative Genome. Sci Am 292: 58-65.

12. Turner BM. 2000. Histone acetylation and an epigenetic code. BioEssays 22: 836-845.
13. Turner BM. 2002. Cellular Memory and the Histone Code. Cell 111: 285-291.
14. Jenuwein T, Allis CD. 2001. Translating the histone code. Science 293: 1074-1080.
15. Brett D, Pospisil H, Valcarcel J, Reich J, Bork P. 2002. Alternative splicing and genome complexity. Nature Genetics 30: 29-30.
16. Xu Q, Modrek B, Lee C. 2002. Genome-wide detection of tissue-specific alternative splicing in the human transcriptome. Nucleic Acids Research 30: 3754-3766.
17. Jaenisch R, Bird A. 2003. Epigenetic regulation of gene expression: how the genome integrates intrinsic and environmental signals. Nature Genetics 33: 245-254.
18. True, H., Berlin I, Lindquist SL. 2004. Epigenetic regulation of translation reveals hidden genetic variation to produce complex traits. Nature 431: 184-187.
19. Wang Y, Fischle W, Cheung W, Jacobs S, Khorasanizadeh S, Allis CD. 2004. Beyond the Double Helix: Writing and Reading the Histone Code. In: G. Bock and J. Goode (eds), Reversible Protein Acetylation. Novartis Foundation 2004, 259: 3-17.
20. Jablonka E, Lamb MJ. 1989. The inheritance of acquired epigenetic variations. J Theor Biol 139 :69-83.
21. Jablonka E, Lamb MJ. 2002. The changing concept of Epigenetics. In: L.v. Speybroeck, G.V.d.Vijver, D.D.Waele, (eds). From Epigenesis to Epigenetics. The Genome in Context. Ann NY Acad Sci 981: 82-96.
22. Jablonka E, Lamb MJ. 2006. The evolution of information in the major transitions. J Theor Biol 239: 236-246.
23. Van De Vijver G, Van Speybroeck L, De Waele D. 2002. Epigentics: A Challenge for Genetics, Evolution and Development. In: L.v. Speybroeck, G.V.d.Vijver and D.DeWaele (eds). From Epigenesis to Epigenetics. The Genome in Context. Ann NY Acad Sci 981: 1-6.
24. Van Speybroeck L, Van de Vijver G, DeWaele D. 2002. Preface, In: L.v. Speybroeck, G.V.d.Vijver and D. De Waele (eds). From Epigenesis to Epigenetics. The Genome in Context. Ann NY Acad Sci 981: vii

25. Witzany G. 1995. From the 'Logic of the Molecular Syntax' to Molecular Pragmatism. Explanatory deficits in Manfred Eigen's concept of language and communication. Evol Cog 1: 148-168.
26. Witzany G. 2000. Life: The Communicative Structure. A New Philosophy of Biology. Norderstedt: Libri Books on Demand.
27. Witzany G. 2005. Natural history of life: History of communication logics and dynamics. S.E.E.D. J 5: 27-55.
28. Mattick JS. 2001. Non-coding RNAs: the architects of eukaryotic complexity. EMBO Rep 2: 986-991.
29. Mattick JS. 2003. Challenging the dogma: the hidden layer of noncoding RNAs in complex organisms. BioEssays 25: 930-939.
30. Mattick JS. 2005. Das verkannte Genom-Programm. Spektrum der Wissenschaft 3: 62-69.
31. Mattick JS, Gagen MJ. 2001. The Evolution of Controlled Multitasked Gene Networks: The Role of Introns and Other Noncoding RNAs in the Development of Complex Organisms. Mol Biol Evol 18: 1611-1630.
32. Watson JD, Witkowski J, Gilman M, Zoller M. ^{2}1992. Recombinant DNA. Scientific American Books.
33. Mantegna RN, Buldyrev SV, Goldberger AL, Havlin S, Peng CK, Simons M, Stanley HE. 1994. Linguistic Features of Noncoding DNA Sequences. Phys Rev Let 73: 3169-3172.
34. Vendrami D. 2004. Noncoding DNA and the teem theory of inheritance, emotions and innate behaviour. Med Hyp64: 512-519.
35. Carroll SB. 2005. Evolution at Two Levels: On Genes and Form. PLoS Biology 3(7): e245.
36. Ryan FP. 2006. Genomic creativity and natural selection. A modern synthesis. Biol J Linn Soc 88: 655-672.
37. Villarreal LP. 2005. Viruses and the Evolution of Life. Washington: American Society for Microbiology Press.
38. Wittgenstein L. 1972. Philosophical Investigations. Oxford: Basil & Blackwell.
39. Villarreal LP. 2004. Can Viruses Make Us Humans? Proc Am PhilSoc 148: 296-323.

40. Villarreal LP, DeFilippis VR, Gottlieb KA. 2000. Acute and Persistent Viral Life Strategies and Their Relationship to Emerging Diseases. Virology 272: 1-6.
41. Villarreal LP, DeFilippis VR. 2000. A hypothesis for DNA viruses as the origin of eukaryotic replication proteins. J Virol 74: 7079-7084.
42. Gao X, Havecker ER, Baranov PV, Atkins JF, Voytas DF. 2003. Translational recoding signals between *gag* and *pol* in diverse LTR retrotransposons. RNA 9: 1422-1430.
43. Forterre P. 2001. Genomics and early cellular evolution. The origin of the DNA world. Comptes rendus de l'Académie des sciences. Série 3, Sciences de la vie 324: 1067-1076.
44. Forterre P. 2002. The origin of DNA genomes and DNA replication proteins. Curr Op Microbiol 5: 525-532.
45. Forterre P. 2005. The two ages of the RNA world, and the transition to the DNA world: a story of viruses and cells. Biochimie 87: 793-803.
46. Forterre P. 2006. The origin of viruses and their possible roles in major evolutionary transitions. Vir Res 117: 5-16.
47. Koonin EV, Senkevich TG, Dolja VV. 2006. The ancient virus world and evolution of cells. Biology Direct 1: 29.
48. Roossinck MJ. 2005. Symbiosis versus competition in plant virus evolution. Nat Rev Microbiol 3: 917-924.
49. Fire A. 2005. Nucleic acid structure and intracellular immunity: some recent ideas from the world of RNAi. Quart Rev Bioph 38: 303-309.
50. Buchon N, Vaury C. 2006. RNAi: a defensive RNA-silencing against viruses and transposable elements. Heredity 96: 195-202.
51. Gabora L. 2006. Self-other organization: Why early life did not evolve through natural selection. J Theor Biol241: 443-450.
52. Margulis L. 1996. Archaeal-eubacterial mergers in the origin of Eukarya: Phylogenetic classification of life. Proc Natl Acad Sci 93: 1071-1076.
53. Margulis L. 1999. Die andere Evolution. Heidelberg: Spektrum Akademischer Verlag.
54. Margulis L. 2004. Serial endosymbiotic theory (SET) and composite individuality. Transition from bacterial to eukaryotic genomes. Microbiol Tod 31: 173-174.

55. Margulis L, Dolan MF, Guerrero R. 2000. The chimeric eukaryote: origin of the nucleus from the karyomastigont in amitochondriate protists. Proc Natl Acad Sci 97: 6954-6959.
56. Margulis L, Sagan D. 2002. Acquiring Genomes. A Theory of the Origin of Species. New York: Basic Books.
57. Witzany G. 2006 a. The Biosemiotic Update. Acta Biotheor 54: 103-117.
58. Odintsova MS, Yurina NP. 2000. RNA Editing in Plant Chlorplasts and Mitochondria. Fisiologia Rastenij 37: 307-320.
59. Odintsova MS, Yurina NP. 2005. Genomics and Evolution of Cellular Organelles. Russ J Gen 41: 957-967.
60. Bell PJL. 2001. Viral Eukaryogenesis: Was the ancestor of the nucleus a complex DNAVirus? J Mol Evol 53: 251-256.
61. Bell PJL. 2006. Sex and the eukaryotioc cell cycle is consistent with a viral ancestry for the eukaryotic nucleus. J Theor Biol doi: 10.1016/j.jtbi.2006.05.015
62. Batzer MA, Deininger DL. 2002. ALU Repeats and Human Genomic Diversity. Nat Rev Gen 3: 370-380.
63. Ryan FP. 2004. Human endogenous retroviruses in health and disease: a symbiotic perspective. J Roy Soc Med 97: 560-565.
64. Dupressoir A., Marceau G, Vernochet C, Benit L, Kanellopoulos C, Sapin V, Heidmann T. 2005. Syncytin-A and syncytin-B, two fusogenic placenta-specific murine envelope genes of retroviral origin conserved in Muridae. Proc Natl Acad Sci 102: 725-730.
65. Eigen M, Schuster P, Gardiner W, Winkler-Oswatitsch R. 1981. The origin of genetic information. Scientific American 244: 78-94.
66. Maynard Smith J. 1983. Models of Evolution. Proc Roy Soc Biol Sci 219: 315-325.
67. Wächtershäuser G. 1992. Groundworks for an evolutionary biochemistry: the iron-sulphur world. Progr Biophys Mol Biol 58: 85-201.
68. Jortner J. 2006. Conditions for the emergence of life on the early earth: summary and reflections. Philos Transac Roy Soc 361: 1877-1891.
69. Deamer D, Singaram S, Rajamani S, Kompanichenko V, Guggenheim S. 2006. Self-assembly processes in the prebiotic environment. Philos Transac Roy Soc 361: 1809-1818.

70. Szathmary E. 2006. The origin of replicators and reproducers. Philos Transac Roy Soc 361: 1761-1776.
71. Witzany G. 2006 b. The *Logos* of the *Bios* 1. Contributions to the Foundation of a three-leveled Biosemiotics. Helsinki: Umweb.
72. Vetsigian K, Woese C, Goldenfeld N. 2006. Collective evolution and the genetic code. Proc Natl Acad Sci 103: 10696-10701.
73. Taft RJ, Mattick JS. 2004. Increasing biological complexity is positively correlated with the relative genome-wide expansion of non-protein-coding DNA sequences. Genome Biol 5: 1.
74. Eigen M, Winkler R. 1975. Das Spiel – Naturgesetze steuern den Zufall. München: Piper.
75. Zhang HY. 2006. The evolution of genomes and language. EMBO Rep 7:248-249.

BOX 1:

Examples of viral genome editing by inventing new genes as shown by Villarreal:

- ✓ RNA, DNA
- ✓ replicase, polymerase, integrase
- ✓ DNA repair
- ✓ restriction / modification
- ✓ methylation
- ✓ bilayer nuclear envelope
- ✓ eukaryotic nucleus
- ✓ division of transcription and translation
- ✓ nuclear pores
- ✓ tubulin-based chromosome duplication
- ✓ chitin, calcification
- ✓ linear chromosomes
- ✓ innate immune system (MHC-Komplex, RNAi)
- ✓ adaptive immune system
- ✓ cartilage, bones
- ✓ skin, dermal glands for poison, mucus and milk
- ✓ larvae, egg, placenta, flowering plants
- ✓ viviparous mammals

6

Serial Endosymbiotic Theory (SET): The biosemiotic Update*

Abstract: The Serial Endosymbiotic Theory explains the origin of nucleated eukaryotic cells by a merging of archaebacterial and eubacterial cells. The paradigmatic change is that the driving force behind evolution is not ramification but merging. Lynn Margulis describes the symbiogenetic processes in the language of mechanistic biology in such terms as "merging", "fusion", and "incorporation". Biosemiotics argues that all cell-cell interactions are (rule-governed) sign-mediated interactions, i.e., communication processes. As the description of plant communication demonstrates, the biosemiotic approach is not limited to the level of molecular biology, but is also helpful in examining all sign-mediated interactions between organisms on the phenotypic level. If biosemiotics also uses the notions of "language" and "communication" to describe non-human sign-mediated interactions, then the underlying scientific justification of such usage should be critically considered. Therefore, I summarize the history of this discussion held between 1920 and 1980 and present its result, the pragmatic turn.

Keywords: symbiogenesis, biosemiotics, genome editing competence, pragmatic turn, abstractive fallacy

---

* Acta Biotheoretica (2006) 54(1).

1. Introduction

The Serial Endosymbiotic Theory explains the origin of nucleated eukaryotic cells by a merging of archaebacterial and eubacterial cells in anaerobic symbiosis, historically followed by acquisition of mitochondria or plastids. Thus, in contrast to former evolutionary theories which consider ramification as a driving force of evolution, there is a paradigmatic change bringing merging into the forefront of the discussion.

Lynn Margulis describes the symbiogenetic processes in the language of mechanistic biology in such terms as "merging", "fusion", "incorporation", and "amalgamation". Biosemiotics argues that all cell-cell interactions are rule-governed sign-mediated interactions, i.e., communication processes. The possibility to investigate plant communication demonstrates that the biosemiotic approach is not limited to the level of molecular biology, but is also helpful in examining all rule-governed sign-mediated interactions between organisms on the phenotypic level.

If biosemiotics also uses the notions of "language" and "communication" to describe non-human sign-mediated interactions, then the underlying scientific justification of such usage should be critically considered. Therefore, I summarize the history of this discussion held between 1920 and 1980 and present its result, the pragmatic turn. Pragmatic philosophy of biology, the pragmatic turn within it, has laid the foundation of a three-leveled biosemiotics and has justified it. It helps to avoid an abstractive fallacy – recurring weakness of today's scientific methodologies.

To extend the Serial Endosymbiotic Theory through a biosemiotic perspective I will follow eight steps as outlined in the following:

The first step is that I briefly present the aims of biosemiotics. Biosemiotics investigates life processes, and more concretely, sign use within and between organisms. Biosemiotics holds that no life processes could be coordinated and organized without signs.

The second step is the outline of the Serial Endosymbiotic Theory. As Lynn Margulis plausibly demonstrates, chance mutations were not responsible for initiating the evolutionary process from prokaryotes to eukaryotes. Rather, merging processes by various prokaryotes – via endosymbioses and the fusion of different organisms – ultimately led to a eukaryotic cell.

The third step allows me to show that these merging processes go beyond mere changes in the chemical/physical state of these cells. Lynn Margulis uses her terms to describe these processes: merging, fusion, incorporation, amalgamation. However, I argue that the symbiogenetic integration of different genomes into a single genome is not a chance merging or fusion process, but rather it requires precise and error-free DNA/RNA processing, not the chance integration of DNA components, but rather the integration of entire gene blocks along with their phenotypic features.

The fourth step shows that competent DNA-processing underlies semiotic rules. Traditionally we have assumed that such nucleic acid processing could only be carried out by enzyme proteins: they were known to be responsible for key DNA modification processes such as DNA splicing, RNA editing, coding, copying, major and fine repair, transcription, translation, etc. They were considered unlikely candidates for evolutionary genome editing. This would have created a paradox that there had to be a DNA that coded for proteins that changed their own DNA-template, as Watson formulated ("Enzymes cannot determine the order of the amino acids in polypeptide chains".)[1] Here, I show that this genome-editing competence, with its higher-order regulatory function, was found not in the protein-coding DNA (as it was previously thought), but in the non-protein-coding DNA (long-termed insignificant "junk DNA").

The fifth step covers the example with plants where I show that the biosemiotic thesis, namely that sign-mediated interactions are the central aspect in coordinating and organizing life processes, is valid not only in the intracellular-genetic realm but also in the intercellular-organismic realm.

The sixth step demonstrates the necessity to specify the scientific basis for referring to language and communication in living nature. The history of science shows that the theoretical clarification of these two terms had two historical phases. The first was the linguistic turn and attempted to construct a language that could be applied to formulate exact statements in the natural sciences. During the second phase, the pragmatic turn, it became apparent that the attempt to justify such a language failed because the proponents had based their premises on an unrealistic model of language and the subject of knowledge.

The seventh step is my applying of a biosemiotic approach to biological processes (e.g. molecular biology, symbiogenesis, or phenotypic communication within and between plants) which requires considering all three semiotic levels of rules (pragmatics, syntax, semantics). A reductionistic abbreviation of these three levels irrevocably leads to an abstractive fallacy, for which there are numerous examples in the history of science.

The eighth step is an attempt to symbiologically expand the self-awareness of biologists. Pragmatic philosophy of biology could help to alter the process of self-awareness in biologists: they should comprehend themselves as biological entities who are permanently involved in symbiotic processes; even in scientific discourse, they are dependent participants in the indefinite dialogue within the scientific community.

2. What's the aim of Biosemiotics

Biosemiotics is a transdisciplinary science including theoretical and empirical studies; it investigates the use of signs within and between organisms. Signs may be signals or symbols, most of them chemical molecules. In the highly developed eukaryotic kingdoms, behavioral patterns of organisms also may serve as signs (signals and/or symbols), as for example, the dances of bees; or the signs may be phonetic, as in songbirds or humans. The signs obey semiotic rules of three types. Syntactic rules determine combinatory possibilities – physical, chemical, spatial, temporal, rhythmical. Pragmatic rules determine interactional content (e.g., regulatory pathways have significant differences with metabolic pathways). And those rules dependent on pragmatic interactional content are the semantic ones, i.e., the meaning of signs and sign sequences (e.g. in signalling pathways).

Individuals in a population share a common set of signs and a common set of rules. It can be applied to cell biology as well. Dependent on the situational context of interacting entities, one sign, or sequence of signs, can have different meanings or functions. Therefore, it is possible that different cell types come from the same genome by the interpretation of different chromosomal methylation patterns. Biosemiotics in the context of molecular cell biology includes not only sign processes used within cells, but also embraces immunological, metabolic, neurological and hormonal signaling pathways.

Zoosemiotics is considered the forerunner of contemporary biosemiotics. Corresponding to the five taxonomic kingdoms zoosemiotics has been completed by prokaryotic semiotics (bacteria), protoctist semiotics (eukaryotic microorganisms), mycosemiotics (fungi) and phytosemiotics (plants).[2] To many biosemioticians, the origin of life is the starting point of semiosis and vice versa.[3-8]

So far, biosemiotic terms have been used as metaphors in molecular and evolutionary biology, as well as in genetics and ecology, the conviction being that they could ultimately be replaced by chemical and physical descriptions. As a result, the paradigmatic differences between biosemiotic and chemical/physical descriptions are becoming much more evident, what enables biosemiotics to draw a clear distinction between the biotic and abiotic domains: "Life is distinguished from the nonliving world by its dependence on signs".[8] Thereby, it is possible to use biosemiotics to expand the perspective on biological processes. Consequently, the factors decisive for life processes do not consist only of the states of matter and their corresponding changes based on natural laws, but also of the communication and information processes within and among cells, tissues, organs, organisms. Their importance determines the success or failure to promote life, growth, development, disease and death in all living beings. Biosemiotics enables a broader understanding of the complexity of life processes which could hardly be achieved by conventional biological methods.

3. The Serial Endosymbiotic Theory

The Serial Endosymbiotic Theory is so revolutionary because it reversed the evolution vector from ramification to merging. Eukaryotic cells, according to Margulis, are the result of merging several different ancestor genomes.[9-13]

Margulis refers to the term *symbiogenesis* as coined by Mereschkovsky and Wallin. The authors argue that new tissues, organs, organisms and species arise by entering into long-term, permanent symbioses. Bacteria were permanently incorporated into animal and plant cells as plastids and mitochondria.[14]

The important factor is the sequence of merging in symbiogenesis, i.e. the serial evolution. The first merger involved (1) thermoplasmic archaebacteria with motile spirochaeta-like eubacteria that (2) were able to swim, to the nucleocytoplasm. These forms were still anaerobic. This was followed by a merging with (3) aerobic organisms. This enabled them to survive the increasing oxygen concentrations. The late and final step was the merging with (4) photosynthetic bacteria. With this approach, the Serial Endosymbiotic Theory contradicts traditional theories of evolution, all of which firmly held that the direction lay in ramification and not in merging.

Cilia, the rods in the retina, the tail of sperm cells, sensory hairs and many similar appendages of nucleated cells arose during the original merging of archaebacteria and a swimming bacterium. The bodies they contain so-called "centriole-kinetosomes" arose through this archaic merger. There is also a genetic relationship between cilia and microtubules at the surface of nuclei of plants, and between microtubules and the mitotic spindles responsible for chromosome movement during cell division. Baluska et al.[15] have convincingly reconstructed this in their *cell body*-theory. According to Margulis, the Spirochaeta are the modern, free-living relatives of these symbiogenetically merged centriole-kinetosomes.

One integrated genome was sufficient in the merger of archae- and eubacterium, in the Proctista two integrated genomes were necessary, in the fungi three, in animals at least four, and in the plant kingdom (400 million years ago) at least 5, perhaps even 7.[9] Thus, the plant genome is the epitome of symbiogenetic evolution processes and represents the most complex integration process. Considering that the evolution of flowering plants took place only about 150 million years ago, and that their seeds and fruits provided the foundation for higher animals, then these revolutionary symbiogenetic processes are relatively young compared with evolutionary history as such.[16]

The Serial Endosymbiotic Theory also supports her assumption by pointing out that most of the DNA found in the cytoplasm of animals, fungi, plants and protoctists comes from genes of bacteria that became organelles, and not from genetic drift or mutations. Eukaryotic genes that participate in information processing (translation, transcription etc.) show a close relationship to eubacteria. Genetic factors that control metabolic processes, however, more closely resemble those of archaebacteria.

Another advantage of the merging paradigm of the Serial Endosymbiotic Theory is that DNA elongation (from bacterium to humans: 1 – 1000 mm) need not be attributed to chance, which would be highly improbable,[17] but rather to a merging of entire gene-blocks. This demonstrates that complex genomic make-ups can be passed on directly and that the step-by-step development via chance mutations is outdated. There are strong reasons too, that eukaryotic nucleus is of viral origin. Endogenous retroviruses are held to be competent in creating new genes in large numbers, complex and simple ones.[18-21]

4. Merging? – Communication!

Lynn Margulis uses in the Serial Endosymbiotic Theory the classic language of mechanistic biology to describe the highly complex interactions of a symbiosis and, subsequently, symbiogenesis. This language ("merging", "fusion", "incorporation", etc.) is imprecise because it describes the altered state of matter, rather than the semiotic aspects of genome editing, i.e., the syntactic, pragmatic and semantic rules which determine combinatorial, interactional and functional aspects of the integration processes of different prokaryotic genomes into one.

In fact, a multi-leveled, generative DNA-processing rather than "fusion" is involved. Its success, however, depends on whether the necessary sign processes proceed according to rules or whether they fail. Moreover, the integration of genetic components into available genomes and, therefore, the creation of a new individual do not involve "incorporation", but rather rule-governed sign-mediated interactions between cells and cell components.

The pragmatic philosophy of biology[3,22] and the young scienceof biosemiotics[23] demonstrate that life functions are always related to sign processes. More precisely, sign processes regulate and constitute life functions.[3,24] If these sign processes are deformed or defective, then life processes are compromised or terminated. Sign processes regulate life processes on different levels simultaneously (Figure 1): intracellularly, within the cell (DNA, RNA activities, messenger substances, etc.), and intercellularly as cell-cell communication. This is the intraorganismic level. In parallel, each organism also conducts (speciesspecific) interorganismic and (trans-specific) metaorganismic communication processes.[3,22]

Should the symbiosis lead to a symbiogenesis, to the development of a new species and, thus, to the disappearance of the formerly independent individuals, then the result is generative DNA- processing in which genetically different gene pools are combined into one genome. This requires a recombination that assimilates the foreign data set, converting the external into the internal. Which genome editing competences are able to integrate an endosymbiotic genome in a host genome? Manfred Eigen would ask how we should think about the correct rearrangement of the "molecular syntax".[25]

5. Symbiogenesis by DNA/RNA processing

Over the last 25 years, tens of thousands of papers have been published in the field of molecular biology, genetics, biochemistry, epigenetics and similar disciplines. They outline in great detail the intracellular processes of recombinant DNA, splicing, RNA-editing, coding, copying, major and fine repairs, transcription, translation, RNA processing, insertion, the role of introns and exons in "reading" processes, the complementary roles of DNAs and RNAs, even the significance and indispensable structural function of non-coding DNA.[26-32,15]

Successful DNA/RNA processing requires numerous, specifically tailored enzyme proteins. In all cases, DNA/RNA-processing enzyme proteins and also interacting RNAs are involved in very precisely conducting these varied DNA - processing steps. Any mistakes here typically have grave and often lethal consequences for the organism.

Today there are strong reasons, that this processing on protein-coding DNA is overruled by the genome processing abilities of DNA coding not for proteins but for active micro-RNAs.[33-38,29] Especially the recombination of two different genomes into one as happened in symbiogenesis we can imagine through the text processing competences of active micro RNAs.

Active micro-RNAs control and integrate large-scale structures of the chromosome. The number of different micro-RNAs is estimated to exceed several 10 000. Some of the discovered tasks of these micro-RNAs are co-suppression, suppression of transposition, position effect variegation, start-stop signals, RNA interference, imprinting, chromosomal methylation, transvection, transcriptional and post-transscriptional gene silencing along with numerous other RNA-DNA, RNA-RNA (trans-acting RNAs), RNA-protein interactions. These active RNAs are as competent as proteins in catalysing, signaling and switching.

Cellular differentiation and phenotypic variation results primarily from variations in this high order regulation, not in the proteins themselves or in their mutations. The phenotypic variation in complex organisms is the result of a *different use of a set of protein-coding core components.* The higher order regulation in non-protein-coding genome architecture is able to manage a larger genetic data set in its phenotypic range. As far as evolutionary processes are concerned, it is naturally much simpler to change or expand a number of very small control sequences than to duplicate an entire network of protein-coding DNA[27,29] Variations of this higher order regulation can create an enormous spectrum of different protein expression profiles and we can understand why one and the same gene can be used for multiple protein meanings. New reports suggest that the capabilities of non-coding DNA with higher-order regulatory functions descended from ancestral viral genome-editing competences that had been integrated by endogenous retroviruses. While complete genetic functions were transferred to the nucleus after the persistent retroviral infections, the regulatory LTR's (Long Terminal Repeats) which flank the protein-coding genetic dataset are integrated in non-protein-coding regions. In the human genome, several thousand such events are documented.[19-21]

6. Plant multilevel communication competence

Biosemiotic aspects are not only interesting on the intracellular genetic and epigenetic level but also in analyzing sign-mediated interactions between cells of organisms, and between organismic bodies, e.g. plants. Plant scientists formerly thought of plants also in terms of mechanistic biology as automatons. Meanwhile research into the multilevel communication of plants revealed attributes like learning, memory, individuality and plasticity as an expression of so-called "plant intelligence". [39-41] During last 5 years plant research has also shown that the old dichotomy of chemical versus neuronal-electric communication is misinterpretation. Today we know that 99% of neuronal communication is based on chemical messenger substances, and that electric action potentials serve merely to maintain the transport of messenger substances by long neuronal tracts[39] Therefore, the catchword for new directions in plant biology is "plant neurobiology", not "plant physiology".

Plants represent a major success story in evolution and are the most recent organismic kingdom. Higher plants make up 99% of the eukaryotic biomass on our planet. At the same time, this success story also reflects the success of multilevel communicative actions by plants in their intra-, inter- and metaorganismic stages: it represents a crucial dependency on successful communication with microbial communities,[42,43] with fungi (especially in the rhizosphere), with animals (especially with insects), and, in parallel, the multilevel communication processes in and between cells, tissues and the whole body.[39]

The communication on the molecular level between plant tissue and the plant cells is exceptionally complex and encompasses nucleic acids, oligonucleotides, proteins and peptides, minerals, oxidative signals, gases, hydraulic and mechanical signals, electric signals, fatty acids, oligosaccharides, growth regulators, amino acids, numerous secondary products, simple sugars, and many other as yet unstudied aspects.[44]

7. Language and communication: from *linguistic turn* to *pragmatic turn*

For more than 10 years, most biological disciplines have increasingly been referring to "communication" and "language" in describing and explaining interactions in and between cells, between tissues and organs, whole bodies, organisms, species-specific and trans-species interactions. The influence of a linguistic vocabulary is omnipresent and has become irreplaceable. This calls for an up-to-date definition of "language" and "communication" if we are to avoid using these terms in an uncritical, unreflected or merely metaphorical manner.

In referring to the language of life, to nucleic- and amino acid codes, and to communication in linguistic terminology in order to describe essential life processes, we can rely on an unspoken and uncritical pre-understanding of language and communication, i.e. on metaphysical and/or reductionistic prerequisites: We can say that (1.) we are working in standardized experimental setups and that theoretical preconditions are not very interesting. We can say that (2.) we refer to the world of objects in the language of exact natural science whose validity claim is founded on the laws of the physical world. We might assume that (3.) observed things have a direct empirical significance that need not be further questioned because the laws of physics correspond one-to-one with the material foundations of the linguistically constructive human brain (universal syntax). We might also assume (4.) an overlying meta-system in which human populations represent one of the subsystems that communicates within itself and with co-systems in an information-theoretical framework.[25,45,46]

This suddenly leaves us directly in the midst a 60-year-long theory of science discussion that extended from approximately 1920 to 1980. It consisted of two phases, and its first result was (a) the *linguistic turn,* the second result being (b) the *pragmatic turn.*

(a) The *linguistic turn* was the result of an attempt to delimit the *logic of science* from philosophy and other "unscientific" cognitive methods. In the aftermath of Wittgenstein's "Tractatus logico philosophicus", the "Wiener Kreis" around Carnap, Neurath, Feigl, Waismann, Kraft, Frank, Menger, Gödel, Hahn and in further developmental stages also Russel and Tarski held that no subjective phenomenology, philosophy or similar discipline provided a suitable logic for an exact natural science; rather, only protocol propositions of observations that are reproducible in experimental setups are capable of depicting reality on a one-to-one basis; this is also valid for propositions of a language of theory, that would have to be brought into agreement with these protocol propositions.

What is required is a language that can be formalized, e.g. as logical calculations, algorithms. This language would represent universal syntax that would be universally valid (i) for the things of the external world, (ii) for the physical laws and (iii) for the material reality of the brain of humans speaking in formalizable propositions*.[25] The history of science clearly documents the course of this discussion. Logical empiricism had to abandon its efforts to achieve the ultimate validity claim of a physicalistic universal language. Concepts such as empirical significance, initial and marginal conditions, verification und falsification, but above all the disposition terms, could not be adequately derived. Even the concept of "natural law" was justifiable only under the assumption of an arbitrary experimental design that presupposed a free will. Thus, the strongest, centuries-long argument against free will – the determinism of the material world – principally needed autonomous researchers if the term "natural law" was to be used justifiable in the natural sciences.[25]

(b) In fact, the transition from the linguistic turn to the *pragmatic turn* has already occurred inWittgenstein's "Philosophical Investigations" and in his analysis of obeying (following) rules: "Is what we call 'obeying a rule' something that it would be possible for only one man to do, and to do only once in his life? (. . .) It is not possible that there should have been only one occasion on which someone obeyed a rule. It is not possible that there should have been only one occasion on which a report was made, an order given or understood, and so on – To obey a rule, to make a report, to give an order, to play a game of chess, are customs (uses, institutions). To understand a sentence means to understand a language. To understand a language means to be master of a technique."[47]

* The conviction of the possibility of an exact language of science was so deeply embedded - and this might serve as an example of the valuation of emotions in purportedly emotion-free objectivism – that, while it was refuted and ultimately rejected in the theory of science discourse and in the history of science, it was and continues to be considered valid in many standard sciences, curricula, underlying convictions.

In his analysis of the expression "to obey a rule", Wittgenstein provides proof that the identity of meanings logically depends on the ability to follow intersubjectively valid rules with at least one additional subject; there can be no identical meanings for the lone and lonesome subject. Communication is a form of social action.

Following Wittgenstein's analysis of rules, the theory-of-science discourse derived and justified scientific statements based on an intersubjective communicative concept of language and communication. At the same time, it replaced the solipsistic subject of knowledge of subjectivism and objectivism with the "ultimate opinion" of an "indefinite community of investigators" of Peirce. The content of argumentation developed by (language-game) Wittgenstein, Austin, Searle, Apel, and Habermas proved the intersubjective-communicative character of thought, experience and research (the a priori of the linguistic community) and was therefore able to avoid the omnipresent subject-object dichotomy and its unavoidable consequences, solipsism and objectivism.

Before we can consider the terms "language" and "communication", we must understand the language of discussing these terms. We understand sentences in a language in which we are linguistically competent; we understand sentences in which the speaker presents propositions interconnected with validity claims. Intersubjective validity of scientific knowledge requires therefore linguistic communication of meaning and consensus formation via statements. We do not understand the ontology of natural phenomena or empirical observations, physiological processes, systems, physical principles, but rather sentences and actions that underlie grammatical, semantic and pragmatic rules that we share with the members of a linguistic community. A prerequisite for understanding is therefore a historically evolved social lifeworld, which provides the basis for the historical development of the commonly shared language, i.e., a core set of signs which obey semiotic rules.[48,49] This, however, means that problems with understanding can arise if we are unaware of the grammatical, semantic or pragmatic rules that an uttering individual is following. *

8. A three-leveled semiotics and classical variations of the "abstractive fallacy"

The *pragmatic turn* founded the *intersubjective-communicative character of thought, experience and research* and was therefore able to avoid the omnipresent subject-object dichotomy and its unavoidable consequences, solipsism and objectivism or how Thomas McCarthy characterized the monological (third-person) observer perspective: "The monological approach preordained certain ways of posing the basic problems of

---

* In order to reach an understanding with another speaker and establish an interrelationship, four validity claims must be fulfilled: (1.) An utterance must be *understandable.* If the partner cannot understand the utterance, then he or she cannot answer (respond) appropriately. (2.) The utterance must be *correct*, i.e. the expressions used must be the correct ones to express the situation (normative rightness). (3.) It must be *true* – the expressed situation must correspond with reality (propositional truth), (4.) It must be *sincere*, i.e. be meant in the manner in which it was expressed.

The understanding of intersubjective acts of human communication is directed at three levels, on that of (a) linguistic utterances, (b) actions, and (c) body-embedded expressions. Linguistic utterances have an evident (locutionary) communicatory aspect. Depending on the intention, they can use this grammatically clearly visible structure to mean something different: This represents their not-immediately-evident (illocutionary) force, which prompts those who are addressed to react in one way or another *to one and the same grammatical structure* of an expression. And they are part of an (perlocutionary) action; perlocutionary acts are performed with the intention of producing a further effect.

We do not need third-person observations and experimental studies to understand how understanding functions. We can analyze the ordinary language that we ourselves use, in the first or second person, i.e. as a participant; here, we can find all the elements of linguistic and communicative action. In a first step, we can determine that utterances such as requests, orders, questions, insinuations, accusations, approval, declarations, fabrications, etc. are regulative, imperative, expressive, objectifying, innovative, etc. actions with the intention of (a) establishing a commonly held understanding *about something* and (b) establishing an *intersubjective relationship of action that enables a common, coordinated* action or appropriate division of labor (Witzany 2005 a).

thought and action: subject versus object, reason versus sense, reason versus desire, mind versus body, self versus other, and so on."[50] As opposed to the linguistic turn, the pragmatic turn enables an understanding of human language and communication that is coherent with our communicative experience *and* with our subjective life. Such an understanding of language and communication should allow us to describe the *intersubjective-communicative character of thought, experience and research* in a non-reductionistic manner, as well as to describe the everyday prerequisites for successful communication, namely:

- the simultaneous understanding of identical meanings in two interacting partners, as expressed in successfully coordinated activity
- the differentiation between deep and superficial grammar of a statement along with differentiation between locutionary, illocutionary and perlocutionary speech acts with which the statements are made
- the differentiation between communication-oriented action and strategic manipulation of the communicating partners
- the critical judgment of the validity being claimed when making a particular statement

Only with such an *(universal-)pragmatic* concept of language and communication, one which is neither subjectivistic nor objectivistic-naturalistic, can we sensibly determine whether similar structural features exist in the non-human realm, i.e., interactions within and between organisms which are sign-mediated (signals, symbols).[3,22] In human world as well as in non-human living nature sign-use is not a solus ipse event but needs individuals which share common set of signs and semiotic rules.

It therefore makes little sense to refer to sign use in the communication processes of plant or bacterial lifeworlds, all the while presupposing an objectivististic language and communication concept, and using an empirical methodological ideal to explain the evolution of communicative interactions from the amoeba to humans. This would lead to the dead-end of the solipsistic subject of knowledge and hopeless entanglement in the subject-object dichotomy of objectivism/ physicalism/ naturalism as described above by McCarthy.

The reverse pathway is correct, beginning with humans, and human self-understanding, which must be coherent with the used concepts of "language" and "communication".

The semiotics of Peirce is helpful in this respect. It can provide the irreducible conditions for the appropriate analyses of sign-usage and linguistic communication via a three-leveled semiotics.*

According to Peirce, a sign (1) designates something (2) to an interpreter (3). Semiotics is therefore an irreducible three-leveled relation of syntactic, semantic and pragmatic rules. *Each of the 3 elements of the sign function already presupposes in its function the other two.* According to Peirce, all those who reduce this principally irreducible three-leveled relation to two or one level, have fallen victim to an abstractive fallacy. The most common of these are:[51]

- Signs (1) without the signified (some)thing (2) and without sign interpreter (3): it yields abstraction from the a priori of the linguistic community community (i.e., the intersubjective-

* The decisive change versus Kant's solus-ipse subject of knowledge (and subsequently the objectivism in logical empiricism) is the "community of interpretation" in the "community of investigators". *Scientific knowledge does not exist for a solipsistic subject of knowledget, but only for members of a community of interpreters.* With this, Peirce adheres to the intersubjective-communicative character of thought, experience and research. (This is coherent with Wittgenstein's analyses of "obeying a rule".)

communicative character of thought, experience and research as described above) which reflects on the world and produces propositions with validity claims. Examples are linguistic Platonism, model-Platonism of the logic of science in the linguistic turn: in which the linguistic expressions or the explanatory model are the real (some)thing.

- The sign interpreter (3) without signs (1) and without the signified (some)thing (2): it yields idealism of consciousness: abstraction from the a priori of the linguistic community (Descartes, Kant, Fichte, Hegel, Husserl): that is to say subjective/objective reason is the real (some)thing. Language is only a secondary means of explanation; the external world exists only through reason/rationality.

- Sign interpreter (3) and sign (1) without the signified (some)thing (2): this yields semiotic idealism: sign and sign interpreters are the signified (some)thing. Everything is a sign: Pansemiotics/metaphysical semioticism.

- The signified (some)thing (2) without signs (1) and without sign interpreter (3): It yields realism and materialism, pre-Kantian ontology/metaphysics: Being is a physical- chemical law of the material world. We are part of this particular material and can at best recognize approximately, with our sense organs, partial perspectives of this material being. Sign use and sign interpretation are pre-scientific constructions that lack a correspondence in the material world (illusion).

- The signified (some)thing (2) and the sign interpreter (3) without signs (1): yields psychologism or affection-realism; Leibniz,

Locke, Berkeley, Hume, positivism of the sensory data, Popper: the observational language and theoretical language of the logic of science can be brought into agreement. The material function of the sensory organs adheres to a "universal syntax" that is identical to the laws of physics and chemistry.

- The signified (some)thing (2) and the sign (1) without the sign interpreter (3): yields ontosemantic realism, constructivism, systems theory, information theory. Abstraction from the a priori of the linguistic community yields subjectless, syntactic-semantic phase of a modern science of logic in Carnap, Russel, Frege, (Tractatus-)Wittgenstein, Tarski, Popper; solipsism and realism: each speaker/listener is equipped with the same syntax (depicting material reality), alleviating any need for communication about the structure of language or the structure of the world. Abstraction by the cognizant subject produces scientifically exact statements which depict, in agreement with the laws of physics and chemistry, material reality on a one-to-one basis. Subjects of knowledge only disturb this naturalism and muddy the exact cognitive process.

9. Pragmatic philosophy of biology: Biologists as performative participants

A pragmatic philosophy of biology is a philosophy of biology which integrates the irreducible three-leveled semiotics. This approach avoids the abstractive fallacy and allows us to understand language-like structured and communicatively organized living nature in a non-solipsistic, non-objectivistic manner. It can therefore methodologically anchor the perspective of *performative participants* (first/second person perspective) of a planetary communicating community of living nature.

The pragmatic philosophy of biology enables a clear distinction between life and the non-living. The unbridgable gap between a mechanistic and the communicative concept is that rule-governed sign-mediated interactions are restricted to living individuals (-in-populations) and do not underly natural laws solely. The decisive difference between natural laws and semiotic rules is that every living being underlies natural laws in a strict sense. Semiotic rules may be followed or not, may be changed or not, may be generated or not. The fundamental difference between living nature and non-living nature is that life depends on sign-mediated interactions, i.e. communication processes which obey syntactic, pragmatic and semantic rules, whereas crystallization of water to ice, for example, requires solely natural laws rather than semiotic rules.

References

1. Watson JD, Gilman M, Witkowski J, Zoller M. 21992. Recombinant DNA.W.H. Freeman and Company, New York. p 6.
2. Noeth W. 2000. Handbuch der Semiotik. Metzler, Stuttgart.
3. Witzany G. 1993. Natur der Sprache – Sprache der Natur. Sprachpragmatische Philosophie der Biologie. Koenigshausen & Neumann, Würzburg.
4. Hoffmeyer J. 1996. Signs of meaning in the Universe. Indiana University Press, Bloomington.
5. Barbieri M. 2001. The Organic Codes. The birth of semantic biology. Ancona: PeQuod.
6. Markos A. 2002. Readers of the Book of Life. Oxford University Press, Oxford/New York.
7. Emmeche C, Hoffmeyer J. 2005. Code-duality and the semiotics of nature. J Biosem 1: 27–64.
8. Pattee HH. 2005. The physics and metaphysics of biosemiotics. J Biosem 1: 223–238.

9. Margulis L. 1996. Archaeal-eubacterial mergers in the origin of Eukarya: Phylogenetic classification of life. Proc Natl Acad Sci USA 93: 1071–1076.
10. Margulis L. 1999. Die andere Evolution. Spektrum Akademischer Verlag, Heidelberg.
11. Margulis, L. 2004. Serial endosymbiotic theory (SET) and composite individuality. Transition from bacterial to eukaryotic genomes. Microbiology Today 31: 173–174.
12. Margulis, L., M.F. Dolan and R. Guerrero 2000. The chimeric eukaryote: Origin of the nucleus from the karyomastigont in an amitochondriate protists. Proc Natl Acad Sci USA 97: 6954–6959.
13. Margulis L, Sagan D. 2002. Acquiring genomes. A theory of the origin of species. Basic books, New York.
14. Searcy DG. 2003. Metabolic integration during the evolutionary origin of mitochondria. Cell Res 13: 229–238.
15. Baluska F, Volkmann D, Barlow PW. 2004. Eukaryotic cells and their cell bodies: Cell theory revised. Ann Bot 94: 9–32.
16. Margulis L, Schwartz KV. 1988. Five Kingdoms.W. H. Freeman and Company, New York.
17. Vollmert B. 1985. Das Molekül und das Leben. Rowohlt, Reinbeck.
18. Villarreal LP, DeFilippis VR. 2000.Ahypothesis for DNA viruses as the origin of eukaryotic replication proteins. J Virol 74: 7079–7084.
19. Bell P. 2001. Viral eukaryogenesis: Was the ancestor of the nucleus a complex DNA virus? J Mol Evol 53: 251–256.
20. Ryan F. 2002. Darwin's blind spot: evolution beyond natural selection. Houghton Mifflin
21. Villarreal LP. 2004. Can viruses make us humans? Proc Am Phil Soc 148: 296–323.
22. Witzany G. 2000. Life: The communicative structure. A New Philosophy of Biology. Libri Books on Demand, Norderstedt.
23. Kull K. 2005. A brief history of biosemiotics. J Biosem 1: 1–25.
24. Witzany G. 2006. The *Logos* of the *Bios* 1. Contributions to the foundation of a three-leveled Biosemiotics. Umweb, Helsinki.
25. Witzany G. 1995. From the "logic of the molecular syntax" to molecular pragmatism. Evol Cog 1: 148–168.

26. Cavalier-Smith T, Beaton MJ. 1999. The skeletal function of non-coding DNA: New evidence from ancient cell chimeras. Genetics 106: 3–13.
27. Sternberg Rv. 2002. On the roles of repetitive DNA elements in the context of a unified genomic-epigenetic system. Ann NY Acad Sci 981: 154–188.
28. Jaenisch R, Bird A. 2003. Epigenetic regulation of gene expression: How the genome integrates intrinsic and environmental signals. Nature gen suppl 33: 245–254
29. Shapiro JA, Sternberg Rv. 2005. Why repetitive DNA is essential to genome function. Biol Rev 80: 1–24.
30. Schmitt S, Paro R. 2004. A reason for reading nonsense. Nature 429: 510–511.
31. True HL, Berlin I, Lindquist SL. 2004. Epigenetic regulation of translation reveals hidden genetic variation to produce complex traits. Nature 431: 184–187.
32. Wang Y, Fischle W, Cheung W, Jacobs S, Khorasanizadeh S, Allis CD. 2004. Beyond the double helix: Writing and reading the histone code. In: Bock, G. and Goode, J. (Eds.) Reversible Protein Acetylation. Novartis Foundation, 2004.
33. Mattick JS. 2001. Non-coding RNAs: The architects of eukaryotic complexity. EMBO Rep 2: 986–991.
34. Mattick JS. 2003. Challenging the dogma: The hidden layer of noncoding RNAs in complex organisms. BioEssays 25: 930.
35. Mattick JS. 2005. Das verkannte Genom-Programm. Spek Wiss 3: 62–69.
36. Mattick JS, Gagen MJ. 2001. The evolution of controlled multitasked gene networks: The role of introns and other noncoding RNAs in the development of complex organisms. Mol Biol Evol 18: 1611–1630.
37. Spotswood HT, Turner BM. 2002. An increasingly complex code. J Clin Invest 110: 577–582.
38. Turner BM. 2002. Cellular memory and the histone code. Cell 111: 285–291.
39. Trewavas A. 2003. Aspects of plant intelligence. Ann Bot 92: 1–20.
40. Trewavas A. 2004. Aspects of plant intelligence: An Answer to Firn. Ann Bot93: 353–357.

41. Trewavas, A. 2005. Green plants as intelligent organisms. Tren Plant Sci 10: 413–419.
42. Walker TS. 2003. Root exudation and rhizosphere biology. Plant Physiol 132: 44–51.
43. Bais HP, Park SW, Weir TL, Callaway RM, Vivanco JM. 2004. How-plants communicate using the underground information superhighway. Tren Plant Sci 9: 26–32.
44. Baluska F, Volkmann D, Hlavacka A, Mancuso S, Barlow PW. (Eds.) 2006. Communication in Plants. Springer Verlag, Berlin/Heidelberg.
45. Witzany G. 1998. Explaining and understanding LIFE. Semiotica, 120: 421–438.
46. Witzany G. 2002. Reduction of biological phenomena? Deficits of systems theory and the alternatives. In: Schmitz, W. (Ed.), Sign Processes in Complex Systems, Mouton de Gruyter, Berlin/New York, 303–307.
47. Wittgenstein L. 1972. Philosophical Investigations. Basil & Blackwell, Oxford.
48. Witzany G. 2005a. From biosphere to semiosphere to social lifeworlds. Biology as an understanding social science. TripleC-Cognition, Communication, Cooperation 3: 51–74.
49. Witzany G. 2005b. Natural history of life: History of communication logics and dynamics. S.E.E.D. J 5: 27–55.
50. McCarthy, T. 1984. Translator's introduction. In: Habermas, J., The Theory of Communicative Action 1, p. ix, Beacon Press, Boston. Company, Boston.
51. Apel KO. 1974. Zur Idee einer transzendentalen Sprachpragmatik. Die Dreistelligkeit der Zeichenrelation und die "abstractive fallacy" in den Grundlagen der klassischen Transzendentalphilosophie und der sprachanalytischenWissenschaftslogik. In Simon, J. (Ed.) Aspekte und Probleme der Sprachphilosophie. Verlag Karl Alber, Freiburg/München, pp. 283–326.

7

From *Umwelt* to *Mitwelt*

Natural laws versus rule-governed sign-mediated Interactions[*]

Abstract: Within the last decade, thousands of studies have described communication processes in and between organisms. Pragmatic philosophy of biology views communication processes as rule-governed sign-mediated interactions (rule-governed sign-mediated interactions). As sign-using individuals exhibit a relationship to following or not-following these rules, the rule-governed sign-mediated interactions of living individuals differ fundamentally from cause-and-effect reactions with and between non-living matter, which exclusively underlie natural laws. Umwelt thus becomes a term in investigating physiological influences on organisms that are not components of rule-governed sign-mediated interactions. Mitwelt is a term for the investigation of all rule-governed sign-mediated interactions of organisms. Living organisms are never solus ipse subjects of semioses, but share common sets of rules and signs. Life depends decisively on symbiotic communities. Serial Endosymbiotic Theory proved that evolution of higher eukaryotic superkingdom was a merger of anchestral bacteria. The integration of bacterial genomes into eukaryotic genomes was also a step from analog to symbolic genetic codes. Now we know, that socalled "junk"-DNA" has higher order regulatory functions on genome architecture and protein coding DNA plays only the role of a structural vocabulary.

---

[*] *Semiotica* 1/4 (2006)

Keywords: Mitwelt, epigenetic apriori, symbiotic interdependence, genome editing MetaCode.

1. The supplement of the *Umwelt*-concept with a *Mitwelt*-concept

Within the last decade, the tendency to describe interactions in and between organisms as communication processes has continued to grow. Intra- und interorganismic communication is supplemented by metaorganismic communication that is between members of different species, genera and organismic kingdoms. Exploding research in the field of symbioses proved, that in fact life on our planet is entirely dependent upon functioning symbiotic interaction communities.[1-3] A pragmatic philosophy of biology and a three-leveled biosemiotics found and justify communication processes as rule-governed sign-mediated interactions.[4-9] Sign use by interaction partners simultaneously obeys syntactic, semantic and pragmatic rules in principle.[9,10]

The conditions on the planet's surface *without* living organisms are determined entirely by the sun's energy and the laws of physics and chemistry.[11] A planet *with* living organisms, i.e. with biosemiotic processes deviates considerably from this scheme. The gas composition and temperature will vary in a manner that cannot be predicted exclusively by the laws of physics and chemistry alone. Such a planet contains incompatible gas mixtures and temperatures whose relatively stable balance is actively controlled by organisms.

The interaction between organisms and matter is one in which organisms interpret and structure their abiotic environment according to biological principles. The relationship to other organisms is a communicative one: their mutual behavior underlies changeable rules, within the frame of natural laws, more precisely (semiotic) rules of sign use with which the biological individuals interact, i.e. coordinate and organize. The difference is that organisms cannot develop such a "to follow or not follow"- relationship with natural laws, but underlie them in the strict sense.

The *Umwelt* concept cannot do justice to the fundamental differences between natural laws and rule-governed sign-mediated interactions because this concept reduces rule-governed sign-mediated interactions to the information transmission effect and therefore to a level of coding and decoding via an individual organism's physiological sensory organ. The intersubjective-communicative character of the rule-governed sign-mediated interactions, and therefore the relationship to following or not-following commonly shared rules of sign-use, does not manifest itself here.

In investigating communication processes in and between organisms, it is therefore sensible to supplement the *Umwelt* concept with a *Mitwelt* concept.[10] The Mitwelt concept underlines the difference between the cause-and-effect reactions in natural laws and rule-governed sign-mediated interactions: *Umwelt* thus becomes a helpful term in investigating those physiological influences on organisms that are not components of rule-governed sign-mediated interactions. *Mitwelt* is a term for the integrative investigation of all rule-governed sign-mediated interactions of organisms, i.e. all intra-, inter- and metaorganismic communication processes themselves and their developmental history.

For example, plants maintain two types of membrane domains that resemble the neuronal and immunological synapses of animal cells.[12] The *first* one transports Auxin and allows the highly plastic development of plants. The activity of this developmental plant synapse is modulated by two physical factors, light and gravity. The developmental plant synapse assembles those molecules of non-living matter which are necessary to structure the plants non-biological environment according to biological principles, i.e. molecules become features of signs for an interpreting organism. There, the *Umwelt* term is useful in describing how organisms develop themselves by metabolising within natural laws.

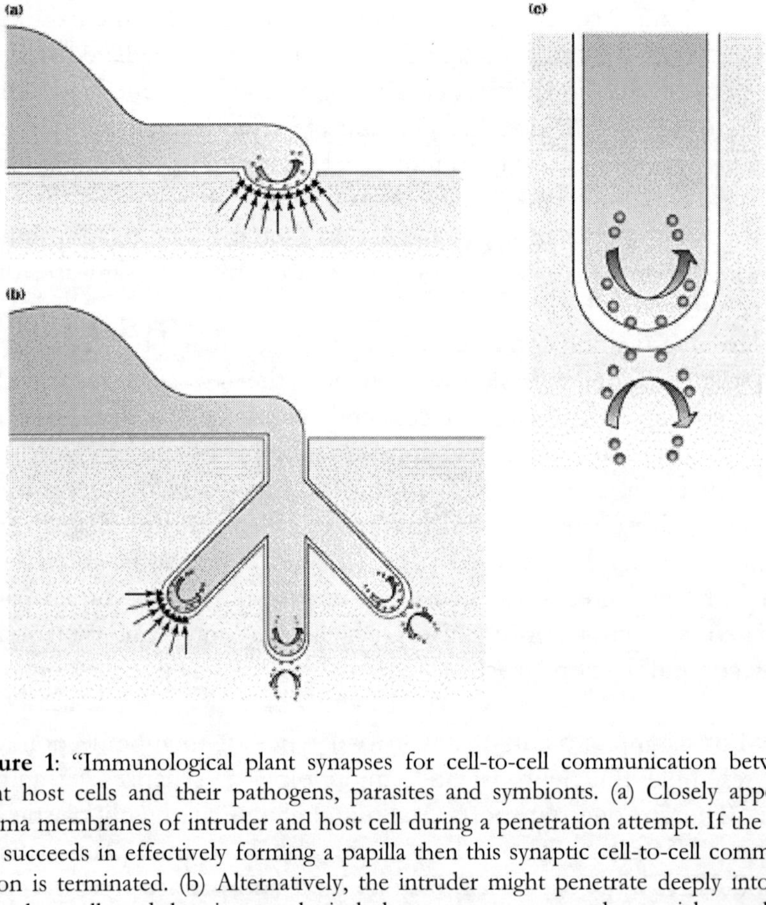

Figure 1: "Immunological plant synapses for cell-to-cell communication between plant host cells and their pathogens, parasites and symbionts. (a) Closely apposed plasma membranes of intruder and host cell during a penetration attempt. If the host cell succeeds in effectively forming a papilla then this synaptic cell-to-cell communication is terminated. (b) Alternatively, the intruder might penetrate deeply into the host plant cells and then immunological plant synapses support haustorial complexes and mycorrhizal arbuscles. (c) During the initiation of a Rhizobia–plant symbiosis, bacteria organize infection threads, the tips of which represent immunological plant synapses specialized for transporting bacteria deeply into root tissues". (figure 1 and text in Baluska et al. 2005: 108).

The *second* type of plant synapse resembles the immunological synapse of animal cells and allows plants to respond to pathogen and parasite attacks[12] as well as to establish stable symbiotic interactions with rhizobia bacteria and fungal mycorrhiza.[12; 13-18]

A limited number of chemical messenger substances is available to maintain and simultaneously conduct the communication between (a) root cells of three different types, (b) root cells and microorganisms, (c) root cells and fungi and (d) root cells and insects.[19-25] The communication process in the root zone is generally intra-, inter- and metaorganismic and requires a high communicative competence in order to be successfully interactive on all three levels and to distinguish messenger molecules from "noise".[26-28] Here, the *Mitwelt* term is useful: its objects are primarily rule-governed sign-mediated interactions, which depend on coherence of signs and three-leveled rules and not on natural laws alone.

2. The epigenetic a priori of semioses

As in the case of semiotics, also biosemiotics depends on the a priori of bodies, i.e. living organisms: without living organisms there are *no* intra- inter and metaorganismic semioses, transcription, translation, copying, coding, splicing, self-splicing (group I and group II introns), i.e., the whole complexity of semiotic processes. Rule-governed sign-mediated interactions depend on living bodies in principle. Living bodies are the *epigenetic a priori of semioses*. Only in rare exceptions living bodies are monads, but they are in any case involved in a commonly shared rule-constituted genetic set up with their descendants.

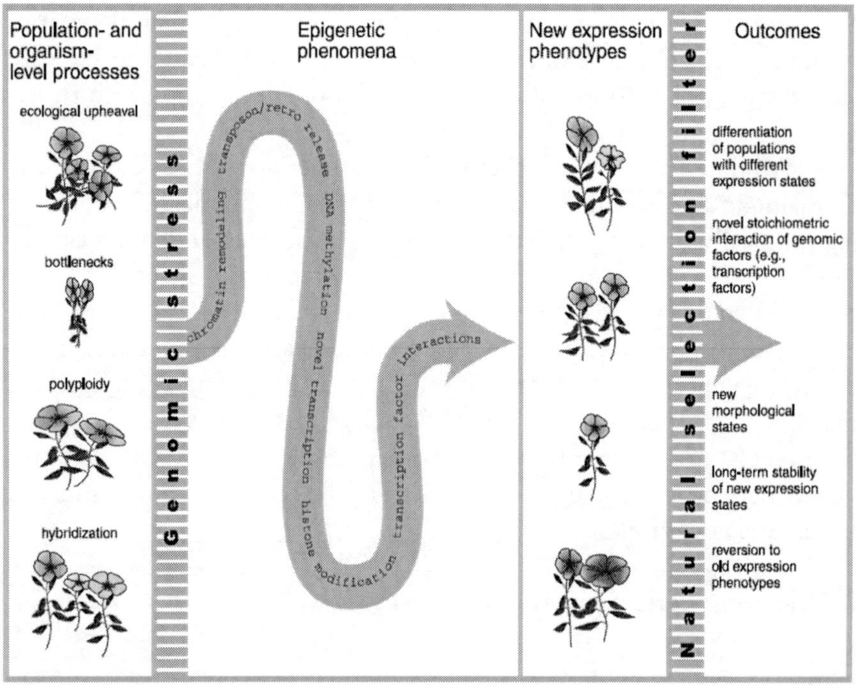

Figure 2: Rapp, R.A. and Wendel, J.F. (2005). Epigenetics and plant evolution. New Phytologist, 168, 81-91.

If signs are used by humans, there is according to Peirce, a non-reducible three leveled relationship between (1) the sign-user/sign-interpreter and (2) the signs which designate (3) (some)thing. According to Morris the relationship: (a) between sign-user/sign-interpreter and signs is subject to pragmatic rules, (b) between signs and the designated (some)thing is subject to semantic rules (c) between signs and signs is subject to syntactic rules. According to Wittgenstein[29] rule obeying sign-users/interpreters are never *solus ipse* subjects (as with Descartes, Kant, Hegel, Husserl, Carnap, Frege, Russel, Tarski and

Popper) but always *performative participants* of communicating communities. Using/interpreting signs is social interacting.

Integrating this pragmatic turn thinkers we are able to avoid the main problem of all variations of philosophy of consciousness: *how to make the move from a state of solus ipse consciousness to a state of mutual agreement, coordination and cooperation.*

Therefore, semiotics as well as biosemiotics are really social sciences.[9] Their research is directed towards signs, their use, their meaning, their syntactic combinations.[30,31] Because one sign alone is not a sign and without a sign-using community there is neither meaning, nor interpretation or linguistic action, i.e. active sign use, the social component is the decisive one. In (language-game-) Wittgenstein it is the analysis of obeying rules, in Peirce it is the "ultimate opinion of an indefinite community of investigators" which refutes the *solus ipse* subject of knowledge of Descartes and Kant in principle.* It is never the role of an interpreter, because "the interpreter" as *solus ipse* subject doesn't exist.[9] Interpreting is a social action, and the "interpretant" is the commonly shared background knowledge of a historically evolved linguistic community.

From this point of reflection we can describe sign use in non-human nature also. But not from an monological *solus ipse* perspective as a cognitive simulation of an (quasi-) extraterrestrial observer, but

---

* The "indefinite community of investigators" of Peirce doesn't resolve the discrepancy between scientific languages and everday language as the ultimate Meta-language. Only the pragmatic action theory was able to found and justify rule-governed sign-mediated interactions as the a priori of the intersubjectiv-communicative character of thought, research, progress of knowledge *and* everyday language. In difference to Peirce the pragmatic action theory doesn't focus on the interpreter of signs in a process of interpretation, but on the *preconditions* for his ability to interpret, i.e. his a priori reliance on a social lifeworld which was the basis for his learning to speak and to act communicatively. In my opinion James Royce with his "community of interpretation" followed by George Herbert Mead and his „community of universal discourse" where more radical in reaching a post-Kantian (communicative) subject of knowledge.

with the knowledge that living bodies are involved in social lifeworlds which depend on social lifeworlds of (a) the same and parallel (b) other organismic kingdoms. All living organisms of the eukaryotic superkingdom *depend decisively* on functioning symbioses with microbial social lifeworlds. In examining the cells of the human body, 15 % stem from the human individual and up to 85 % are symbiotic microorganisms.[32]

3. Evolution of the eukaryotic superkingdom by genomic integration

Another example applying the *Mitwelt*-concept is the Serial Endosymbiotic Theory. The SET is so revolutionary because it reversed the evolution vector from ramification to unification. Eukaryotic cells, according to Margulis, are the result of merging several different ancestor genomes, that of (1) thermoplasmic archaebacteria with (2) motile spirochaeta-like eubacteria. This was followed by a merging with (3) aerobic organisms. The final step was the merging with (4) photosynthetic bacteria. One integrated genome was sufficient in the merger of archae- and eubacterium. In the protoctista two integrated genomes were necessary, in the fungi three, in animals at least four, and in the plant kingdom (350 million years ago) at least five (or even 7).[11,33-39]

From the biosemiotic point of view this symbiogenetic process was a change *from metaorganismic* communication, i.e. rule-governed sign-mediated interactions between two individuals who lived in symbiosis, to an *intraorganismic* communication process. In the latter, two different genomes integrated into one at the correct sequences, at the correct genome ratio, (i.e. the genome architecture). It had to involve a correct integration of entire gene-blocks.

The SET demonstrates that complex genomic make-ups can be passed on directly and not through step-by-step development via chance mutations.[9] One scenario of how this integration process of two different genomes could take place can be imagined by the phenomenon of horizontal gene transfer.[3,40-45] As suggested[9,10,46] these (symbiogenetic and HGT) processes are not changes in aggregate status solely but undoubtable biosemiotic ones: They are highly competent DNA textprocessing procedures.[47-49] Thomas and Nielsen[50] identified 17 steps of integration performance (see Fig. 2).

The DNA transfer between prokaryote cells include transformation, transduction and conjugation which depend on certain mobile genetic elements such as plasmids, bacteriophages and transposons. In this processes that agents are the 'natural genetic engineers'.[47]

4. Analog — symbolic (prokaryotic genome - eukaryotic genome)

The step from prokaryotic to eukaryotic cells and the recombination of up to 5 (or maybe even 7) different genomes into one, as occurred in the evolution of plant cells was also the step from a nearly complete analog proteincoding DNA to a genome architecture of analog and increasingly symbolic sequences.[9] As higher-developed eukaryotic multicellular organisms show, this analog/symbolic-ratio changed into a proportion in humans, for example, of 3 percent proteincoding DNA (which is nearly identical to that of mice, namely 99 percent) and 97 percent non-proteincoding DNA.

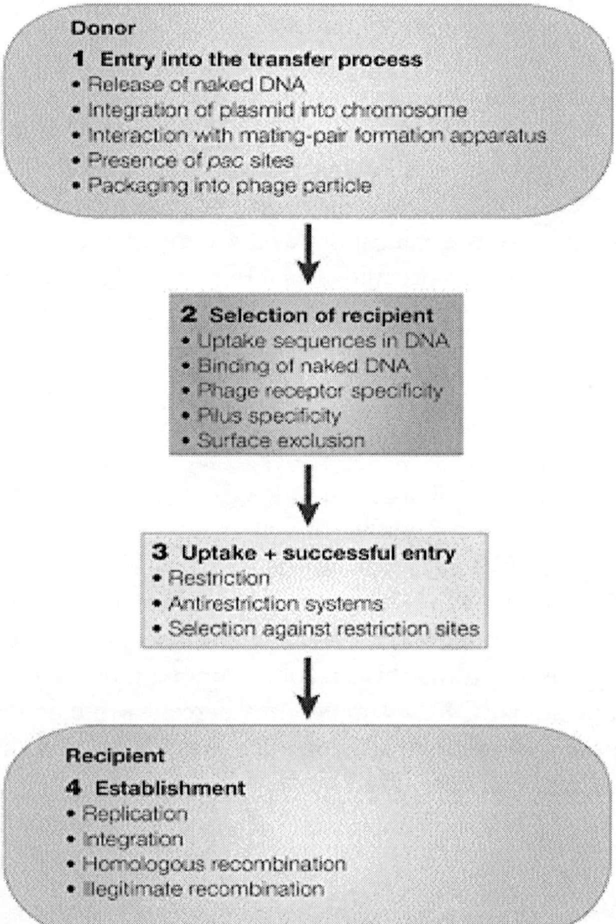

Figure 3: "The process of horizontal gene transfer. A schematic outlining the stages through which DNA must go on its journey from donor to recipient bacteria. The process begins with DNA in a potential donor cell becoming available and ends when this DNA becomes a functional part of a recipient cell's genome." (figure and text in Thomas and Nielsen 2005: 719).

Ten years ago this non-proteincoding DNA was not the center of research interests and therefore named "junk"-DNA. While there is no relation between the complexity of organisms and the quantity of proteincoding DNA, but an obvious relation between complexity of organisms and non-proteincoding DNA, this "junk"-DNA became interesting. It does not code for proteins but for RNAs especially for a type of micro-RNAs.[51-56] These micro-RNAs are as competent as proteins in DNA/RNA-editing. Some may term this discrete structure of eukaryotic DNA (that doesn't indicate analogous for what it codes) digital.[57] But this information theoretical description is a more or less technical term, whereas "symbolic" is a better description for the phenomena as such.[58,9]

Some of the discovered tasks of these micro-RNAs are co-suppression, suppression of transposition, position effect variegation, start-stop signals, RNA interference, imprinting, chromosomal methylation, transvection, transscriptional and posttransscriptional gene silencing along with numerous other RNA-DNA, RNA-RNA (trans-acting RNAs), RNA-protein interactions.[55] Today we may say, that cellular differentiation and phenotypic variation results primarily from variations in this high-order regulation, not in the proteins themselves, thus non-proteincoding DNA bears the architecture of eukaryotic complexity.[52]

5. Structural (protein-code-) vocabulary and its higher-order regulation

The phenotypic variation in complex organisms is then the result of a *different use* of a set of protein-coding core components. We can say that in higher eukaryotic organisms we find 3 percent protein-coding DNA as the structural vocabulary and 97percent non-proteincoding DNA which has higher-order regulatory and constitutional functions that are decisive for expression, differentiation, development, (i.e. coordinated expression in time). I postulated these higher order regulatory and constituting functions as a consequence of the pragmatic philosophy of biology approach already in 1993 and in a further developed version in 2000 in terms like "innovation-code, text-generating-code, evolution-code".[5,59,60]

In an article in S.E.E.D. Journal,[9] I put it in concrete as the genome-editing *MetaCode.* This non-protein coding *geMetaCode* has regulatory and constituting functions in chromosomal methylation. These functions decide about different signalling pathways of the same genetic setups through different rules of gene silencing and special start and stops via alternative splicing. With this chromosomal methylation, organisms are able to handle "multiple protein meanings"[61] of the same genetic dataset. As shown by some epigeneticists, these (*geMetaCode*-) functions are influenced and altered by environmental and other influences which are heritable but *do not involve or depend on mutations* of the DNA itself.[62-70]

The proposed *geMetaCode* integrates the features of (1) non-proteincoding DNA, (2) DNA which codes alternatively for (micro)RNAs and proteins, which depends on situational contexts and (3) the proteincoding DNA. This enables the (epigenetic) body and genome to interact *in time* over the whole lifetime, in the expression pathways of early development, the body axis, extremities, up until the adult individual. Or, as in plants, the never-ending growth and development[71-73] based on *one and the same* protein coding vocabulary. The latter is used depending on the developmental context as epigenetic orientation of the plants, and therefore by the non-proteincoding regulatory, constitutional and generating functions of *geMetaCode.*

6. Genetic backup insurance against mutational faults

Plants can overwrite their genetic code they inherited from their parents and revert to that of their grand or great-grandparents.[74-76] This contradicts traditional DNA-textbook conviction that children simply receive combinations of the genes carried by their parents. Now a backup code was found that can bypass unhealthy sequences inherited from the parents and revert to the healthier sequences possessed by their grandparents or great-grandparents. As shown by a research team plants are able to replace abnormal code sequences of their parents with the regular code possessed by earlier generations.

Is it therefore necessary that not only the genetic set up of the parents is inherited but also that of the grand-parents and former ancestors? What is proposed is that higher order regulating micro RNAs - which function as expressions of *geMetaCode* - inshure ancestor genome structures, which overrule protein-coding DNA under certain circumstances like stress. Such stress situations that could be dangerous for survival, could trigger plants to revert to the genetic setup of their ancestors which perhaps is more competent to deal with this circumstances than that of their parents.

It means that (pragmatic) situational context of a living plant body (neuronal, hormonal, inflammatory, immune) stress, for example, may induce epigenetic intervention on *geMetaCode*, i.e. active micro-RNA's activate a certain signalling pathway network which is able to restructure semantics of a genetic setup. By initiating methylation- and histone- modifications, certain silencings, start and stops, alternative splicing processes constitute alternative sequences, so that in existing genome architecture not the inherited parental sequences are translated and transcribed but the backup copy of grand or great- grandparents.

Under normal conditions genetic set up is in use which stem from the parents. These research results are indications that not only a combination of parental genes is inherited but also ancestral genome-regulating features in "non-coding" DNA, which enables alternative splicing pathways, i.e. a *different use and multiple protein meanings of one and the same genetic data set.*[74-76]

7. Conclusions

This contribution could be an approach to resolve the contradiction of Neo-Darwinism and Neo-Lamarckism: The proteincoding DNA is not subject to evolutionary remodifications that lead to new abilities or new organisms, but is subject to chance mutations which can alter *existing* genotypes. The non-protein coding DNA, with its higher-order regulatory functions is subject to evolutionary remodifications *that are heritable but also reversible* and that regulate the structural vocabulary of protein-coding DNA. The *geMetaCode* is subject of encoding acquired abilities i.e. to higher-order regulatory and constitutional functions, which may reorganize *and generate* new and altered genotypes with phentotypic consequences by alternative use of *one and the same* protein-coding DNA.

References

1. Zook D. 1998. A New Symbiosis Language… . *ISS SymbiosisNews*, 1(3: 1-3.
2. Kowallik KV. 1999. Endosymbiose - ein Motor der Evolution. *biologen heute*, 1.
3. Wagner E. et al. 1999. From Symbiosis to Eukaryotism. ENDOCYTOBIOLOGY VII. *Proceedings of the International Congress on Endocytobiology, Symbiosis and Biomedicine*, University of Freiburg, University of Geneva.
4. Witzany G. 1993a. Zeichenprozesse als Bedingungen der Möglichkeit von Leben und Evolution. *Zeitschrift für Semiotik*, 151/2: 107-125.
5. Witzany G. 1993b. Natur der Sprache – Sprache der Natur. Sprachpragmatische Philosophie der Biologie. Würzburg : Königshausen and Neumann.

6. Witzany G. 1995. From the „logic of the molecular syntax“ to molecular pragmatism. Evol Cog 12: 148-168.
7. Witzany G. 2002a. Reduction of biological phenomena? Deficits of systems theory and the alternatives. In W. Schmitz. Ed. Processes in Complex Systems. Berlin/New York:Mouton de Gruyter, pp 303-307.
8. Witzany G. 2002b. Sprache und Kommunikation als zentrale Struktur- und Organisationsprinzipien belebter Natur. In Albers, L. and O. Leiss. Eds. Körper-Sprache-Weltbild. Integration biologischer und kultureller Interpretationen in der Medizin. Stuttgart/New York: Schattauer, pp 87-96.
9. Witzany G. 2005b. Natural History of Life: history of communication logics and dynamics. SEED J 5: 27-55..
10. Witzany G. 2005 (in press). From biosphere to semiosphere to social lifeworlds. Biology as an understanding social science. J Biosem 2.
11. Margulis L. 1999. Die andere Evolution. Heidelberg: Spektrum Akademischer Verlag.
12. Baluska F, Volkmann D, Menzel D. 2005. Plant synapses: actin-based domains for cell-to-cell communication. Trends Plant Sci 10(3: 106-111.
13. Estabrook EM, Yoder JI. 1998. Plant-plant communications: rhizosphere signalling between parasitic angiosperms and their hosts. Plant Physiol 116: 1-7.
14. Yoder JI. 1999. Parasitic plant responses to host plant signals: a model for subterranean plant-plant interactions. Curr Opin Plant Biol 2: 65-70.
15. Keyes WJ, O'Malley RC, Kim D, Lynn DG. 2000. Signaling Organogenesis in Parasitic Angiosperms: Xenognosin Generation, Perception, and Response. J Plant Growth Regul 192:217-231.
16. Kahmann R, Basse C. 2001. Fungal gene expression during pathogenesis-related development and host plant colonization. Curr Opin Microbiol 4(4):374-380.
17. Engelberth, J, Alborn HT, Schmelz EY, Tumlinson JH. 2004. Airborne signals prime plants against insect herbivore attack. Proc Natl Acad Sci USA 101(6): 1781-1785.
18. Imaizumi-Anraku H, Takeda N, Cherpentier M, Perry J, Miwa H, Umehara Y, Kouchi H, Murakami Y, Mulder L, Vickers K, Pike J, Downie A, Wang T, Sato S, Asamizu E, Tabata S, Yoshikawa M, Murooka Y, Wu

GJ, Kawaguchi M, Kawasaki S, Parniske M, Hayashi M. 2005. Plastid proteins crucial for symbiotic fungal and bacterial entry into plant roots. Nature 433:527-531.

19. Bais HP, Park SW, Weir TL, Callaway RM, Vivanco JM. 2003. How plants communicate using the underground information superhighway. Trends Plant Sci 91: 26-32.
20. Callaway RM. 2002. The detection of neighbors by plants. Trends Ecol Evol 17: 104-105.
21. Dessaux Y. 2004. Biological Communications and Interactions in the Rhizosphere. (unpublished). Presented at „Rhizosphere 2004" Congress, 12.- 17. September, Munich, Germany.
22. Dunn AK, Handelsman J. 2002. Toward an understanding of microbial communities through analysis of communication networks. *Antonie van Leeuwenhoek*, 81: 565–574.
23. Teplitski M, Robinson JB, Bauer WD.2000. Plants secrete substances that mimic bacterial N-acyl homoserine lactone signal activities and affect population density-dependent behaviours in associated bacteria. Mol Plant-Microbe Interact 13: 637-648.
24. Walker TS. 2003. Root exudation and Rhizosphere Biology. Plant Physiol 132: 44-51.
25. Fleming A. (ed.) 2005. Intercellular Communication in Plants. Annu Plant Rev 16.
26. Federle MJ, Bassler BL. 2003. Interspecies communication in bacteria. J Clin Investig 112: 1291-1299.
27. Hirsch AM, Bauer WD, Bird DM, Cullimore J, Tyler B, Yoder JI. 2003. Molecular signals and receptors: controlling rhizosphere interacting between plants and other organisms. Ecology 84: 858-868.
28. Sharma A, Ahgal M, Johri BN. 2003. Microbial communication in the rhizosphere: Operation of quorum sensing. Curr Sci 85(8: 1164-1172.
29. Wittgenstein L. 1968. Philosophical Investigations. Basil & Blackwell, Oxford.
30. Noeth W. 2000. Handbuch der Semiotik. Metzler, Stuttgart.
31. Kull K. 2005. A brief history of Biosemiotics. J Biosem 1: 1-34.
32. Blech J. 2000. Leben auf dem Menschen. Die Geschichte unserer Besiedler. Hamburg: Rowohlt Taschenbuch Verlag.

33. Margulis, L. 1996. Archaeal-eubacterial mergers in the origin of Eukarya: Phylogenetic classification of life. Proc Natl Acad Sci USA, 93: 1071-1076.
34. Margulis L. 2004. Serial endosymbiotic theory (SET) and composite individuality. Transition from bacterial to eukaryotic genomes. Microbiol Today 31: 173-174.
35. Margulis L, Schwartz KV. 1988. Five Kingdoms. New York: W.H. Freeman.
36. Margulis L, Dolan MF, Guerrero R. 2000. The chimeric eukaryote: origin of the nucleus from the karyomastigont in an amitochondriate protists. Proc Natl Acad Sci USA 97: 6954-6959.
37. Margulis L, Sagan D. 2002. Acquiring Genomes. A Theory of the Origin of Species. New York: Basic Books.
38. Cavalier-Smith, T. 2002. Chloroplast evolution: secondary symbiogenesis and multiple losses. Curr Biol 12: 62-64.
39. Stechmann A, Cavalier-Smith T. 2003. The root of the eukaryote tree pinpointed. Curr Biol 1317: 665-666.
40. Wolf YI. 1999. Evolution of Aminoacyl-tRNA Synthetases—Analysis of Unique Domain Architectures and Phylogenetic Trees Reveals a Complex History of Horizontal Gene Transfer Events. Gen Res 9: 689-710.
41. Wolf YI. 2000. Genome Alignment, Evolution of Prokaryotic Genome Organization, and Prediction of Gene Function Using Genomic Context. Gen Res 11: 356-372.
42. Jain R, Rivera MC, Lake JA. 1999. Horizontal gene transfer among genomes: the complexity hypothesis. Proc Natl Acad Sci USA 96: 3801-3806.
43. Jain R, Rivera MC, Moore JE, Lake JA. 2003. Horizontal Gene Transfer Accelerates Genome Innovation and Evolution. Mol Biol Evol 2010: 1598–1602.
44. Xie G, Bonner CA, Song J, Keyhani NO, Jensen RA. 2004. Intergenomic displacement via lateral gene transfer of bacterial trp operons in an overall context of vertical genealogy. MBC Biol 2: 15.
45. Timmis JN, Ayliffe MA, Huang CY, Martin W. 2004. Endosymbiotic gene transfer: organelle genomes forge eukaryotic chromosomes. Nat Rev Gen 5: 123-135.

46. Witzany, G. 2005a. Serial Endosymbiotic Theory (SET): The biosemiotic update 2005. Presented at the 1st Meeting on Plant Neurobiology, Florence, Italy, May 17-20.
47. Frost, LS, Leplae, R, Summers AO, Toussaint A. 2005. Mobile genetic elements: The agents of open source evolution. Nat Rev Microbiol 3: 679-687.
48. Bordenstein SR, Reznikoff WS. 2005. Mobile DNA in obligate intracellular bacteria. Nat Rev Microbiol 3: 688-699.
49. Gogarten JP, Townsend JP. 2005. Horizontal gene transfer, genome innovation and evolution. Nat Rev Microbiol 3: 679-687.
50. Thomas CM, Nielsen KM. 2005. Machanisms of, and barriers to, horizontal gene transfer between bacteria. Nat Rev Microbiol 3:711-721.
51. Cavalier-Smith T. Beaton MJ. 1999. The skeletal function of non-coding DNA: new evidence from ancient cell chimeras. Genetics 106: 3-13.
52. Mattick JS. 2001. Non-coding RNAs: the architects of eukaryotic complexity. EMBO Rep 211: 986-991.
53. Mattick JS. 2003. Challenging the dogma: the hidden layer of noncoding RNAs in complex organisms. *BioEssays*, 2510: 930.
54. Mattick JS. 2005. Das verkannte Genom-Programm. *Spektrum der Wissenschaft*, 3: 62-69.
55. Mattick JS, Gagen MJ. 2001. The Evolution of Controlled Multitasked Gene Networks: The Role of Introns and Other Noncoding RNAs in the Development of Complex Organisms. Mol Bio Evol 18(9: 1611-1630.
56. Shapiro JA, Sternberg Rv. 2005. Why repetitive DNA is essential to genome function. Biol Rev 80: 1-24.
57. Emmeche ¨, Hoffmeyer J. 2005. Code-Duality and the Semiotics of Nature. J Biosem 1: 48
58. Cvrcková F, Markos, A. 2005. Beyond bioinformatics: can similarity be measured in the digital world? J Biosem 1: 87-105.
59. Witzany G. 1997. Semiosis and Evolution. In I. Rauch. Ed. Semiotics around the World: Synthesis in Diversity. Berlin: Mouton de Gruyter, pp 977-980.
60. Witzany G 2000. Life: The Communicative Structure. A New Philosophy of Biology. Norderstedt: Libri Books on Demand.

61. Ast G. 2005. The Alternative Genome. Sci Am 4.
62. Jenuwein, T. and C.D. Allis. 2001. Translating the histone code. Science, 293: 1074-1080.
63. Spotswood HT, Turner BM. 2002. An increasingly complex code. J Clin Invest 110 (5): 577-582.
64. Sternberg Rv. 2002. On the Roles of Repetitive DNA Elements in the Context of a Unified Genomic-Epigenetic System. Ann NY Acad Sci 981: 154-188.
65. Turner BM. 2000. Histone acetylation and an epigenetic code. BioEssays, 22: 836-845.
66. Turner BM. 2002. Cellular Memory and the Histone Code. Cell, 111: 285-291.
67. Jaenisch R, Bird A. 2003. Epigenetic regulation of gene expression: how the genome integrates intrinsic and environmental signals. Nature gen suppl 33: 245-254.
68. True H, Berlin I, Lindquist SL. 2004. Epigenetic regulation of translation reveals hidden genetic variation to produce complex traits. Nature, 431: 184-187.
69. Wang Y, Fischle W, Cheung W, Jacobs S, Khorasanizadeh S, Allis CD. 2004. Beyond the Double Helix: Writing and Reading the Histone Code. In G. Bock and J. Goode. Eds. Reversible Protein Acetylation. Novartis Foundation, 2004.
70. Waterland RA, Jirtle RL. 2004. Early Nutrition, Epigenetic Changes at Transposons and Imprinted Genes, and Enhanced Susceptibility to Adult Chronic Diseases. Nutrition, 201: 63-68.
71. Trewavas A. 2001. How plants learn. Proc Natl Acad Sci USA 96: 4216-4218.
72. Trewavas A. 2003. Aspects of Plant Intelligence. Ann Bot 92: 1-20.
73. Trewavas A. 2005. Green Plants as Intelligent Organisms. Presented at the "First Meeting on Plant Neurobiology", Florence, Italy, May 17-20.
74. Lolle SJ, Victor JL, Young JM, Pruitt RE. 2005. Genome-wide non-mendelian inheritance of extra-genomic information in Arabidopsis. *Nature*, 434: 505-509.
75. Weigl D, Juergens G. 2005. Genetics. Hotheaded Healer. Nature, 434: 443.

76. Pearson H. 2005. Cress overturns textbook genetics. Nature, 434: 351-360.

Epilogue

The contributions of this book are based on empirical data and demonstrate that:

(a) Organisation and coordination within and between cells, tissues, organs, organisms, species-specific and trans-specific (symbiotic) societies require communication processes, i. e., rule-governed sign-mediated interactions.
(b) Biotic agents with genome editing competences from the very beginning of life invent, arrange and rearrange RNA/DNA storage medium.
c) Neither the linguistic competences of the genome editing agents nor the communicative competences of organisms are able to live without the complementary role of the other level.

For decades the interactions within and between organisms have been investigated by the assumption of rather mechanistic stimulus-reaction patterns which in principle can be reconstructed by formalizable procedures (i. e., algorithms) expressed in terms like "machinery", "apparatus", "mechanism", etc. In the last decade it has become undoubtedly clear, that organisms interact by using signals as signs according to combinatorial rules (syntax), contextual rules (pragmatic), and content-specific rules (semantic) which cannot be reduced to one another but are crucial for successful communication processes and cannot be objectivised by formalizable procedures alone.

Organization and coordination of organism behavior depend on successful communication. If one level of rules is deformed, damaged, incomplete, either by the transmitting or by the receiving organism, communication may fail even with fatal consequences for the organisms. First it seemed that communication processes are limited to simple signal exchange but as we become aware that this is the exception and from bacterial communication processes up to vertebrates, primates and plants different levels of rule-governed sign-mediated interactions must function in a parallel and in a continuous way. This includes often the mixed use of different signalling levels in gaseous, aquaeus, cristalyzed or even tactile types.

Biocommunication from the very beginning of life is a type of complex interaction and it opens appropriate behavorial patterns to adapt to nearly all circumstances of planetary biosphere. And it is a precondition to successful adaptation by a rising complexity. In all cases where the use of analog signs to symbols occurred the raise of complexity is manifold, may it be the change of the sequence order from a prokaryotic genome to an eukaryotic one or the different type of behavorial signs by honey bees of the southern or the northern hemisphere.

For more than 50 years it has been assumed that randomly derived (chance) mutations represent the driving force of evolutionary changes of the nucleic acids storage medium as precondition for generation of new organismic phenotypes. We now become aware that this is rather the exception. Genetic and genomic arrangements and rearrangements depend on complex linguistic competences of biotic agents and evolutionary change is accomplished via nearly failure-free nucleic acid text editing.

Parallel to this, it was considered that the evolutionary and developmental phenomena are based exclusively on the DNA - Proteins

- Phenotypes pathway. Now it is obvious that, complementary to this genotype-phenotype direction, there are many other directions. As a consequence, analog sensory patterns of organism phenotypes, mediated by molecular biological pathways, are both, can change phenotypes without alterations of the genetic text simply by changes of regulatory rules (which can be inherited but need not) or are encoded into "digitally" structured nucleic acid storage medium.
This new biosemiotic view on processual reality of most living processes brings some advantages to traditional scholarly conviction of mechanistic perspectives:

- A coherent non-mechanistic and non-reductionistic categorization and description method of biotic interactional patterns.

- A clear distinction between life and non-life.

- Clear terms for equal or similar behavioral patterns of signal generation, signal transduction and signal perception and coherent generation of response behaviour throughout all organismic kingdoms.

- A clear distinction of simulation of interactions in experimental set ups/ models and real live world symbiology.

Biosemiotic perspective on life and its (a) language-like storage medium and (b) communicatively organized organismic interactions could bring advantages in biological research:

- Any observed coordination within and between organisms is proposed to be sign-mediated and can be deciphered by research which identifies signalling molecules and biosemiotic (syntactic, pragmatic, semantic) rules underlying the mode of signal use.

- Research must not concentrate any longer on experimental set-ups which evaluate randomly altered genetic content, because genetic content is not randomly derived.

- The biocommunicative approach with its unique categorization and new perspective on linguistic competences of viruses could avoid scientific, medical and even political fallacies which deal with chance mutations being relevant for pandemic diseases. In almost all observations (chance-) mutations played no role in the emergence of new phenotypes with advantages for (new) species but neutral or disease causing and lethal consequences. Alterations or inventions of genetic/genomic content is nearly in all cases the result of high level linguistic competences in performative genome editing of viral agents.

- Although all biological processes are coherent with laws of physics and chemistry, empirical data show that there is a biotic plus which cannot be deduced from physics and chemistry alone. It is the invention and use of signs according to syntactic, pragmatic and semantic rules. We can observe and describe these rules and, therefore, we are able to understand signals and symbols. This is a type of hermeneutic/semiotic research. Additionally biological processes are interlinked with organisms in real lifeworlds, i.e. in symbiotic/parasitic dependencies. This means that the major methodological paradigm of biology is the one of an understanding social science.

- Biological research must not concentrate any longer on experimental set-ups and theoretical approaches which want to elucidate language and communication in non-human living nature by stochastic, physicalistic, mathematic, systems theoretic modelling, because formalizing approaches cannot investigate non-formalizable rules of content-generation. Thus, the explanatory power of the biocommunitive approach is more coherent with the complexity of living phenomena than formalizing models.

- Even if a LaPlace Demon would exist, with insight into the whole physical status quo, he could not predict the creative output of biological agents as these are equipped with inherent linguistic and communicative competences underlying their inherent non-deterministic features.

„Witzany has successfully developed an independent theory of living nature that opens the door to a comprehensive understanding of living nature."

Thure von Uexküll

Content

1. Introduction
2. From the „logic of the molecular syntax" to molecular pragmatism
3. Semiosis and Evolution
4. Explaining and understanding LIFE
5. Reduction of biological phenomena
6. The origin of novel entities in living nature
7. Metaphysics in physics and biology
8. From biosphere to semiosphere to social lifeworlds
9. Natural history of life: history of communication logics and dynamics

The *Logos* of the *Bios* 1
Contributions to the Foundation of a three-leveled Biosemiotics.
Guenther Witzany,
Helsinki. Umweb, 220 p.

It opens a new perspective on living nature through the philosophical foundation of a biology as an understanding social science. The contributions integrate the pragmatic turn of the theory of science discussion which replace the *solus ipse* subject of knowledge of objectivism by the intersubjective- communicative character of thought, experience and research.

A three-leveled biosemiotics investigates rule-governed sign-midiated interactions within and between organisms of all organismic kingdoms. Therefore it is decisive to keep in mind the complementarity of syntactic, pragmatic and semantic rules as precondition for an appropriate investigation of the language-like structured and communicatively organized living nature.

€ 32,00

(+ forwarding expenses)

online order:
witzany@sbg.at

The Author: Günther Witzany, born 1953 in Salzburg/Austria founded the 1st Philosophische Praxis in Austria in 1985. His research interests are applied philosophy, linguistic pragmatics, biosemiotics, molecular biology, genetics and epigenetics, bacteria-, fungi-, animal and plant communication.